RELATIVISTIC PHYSICS IN ARBITRARY REFERENCE FRAMES

RELATIVISTIC PHYSICS IN ARBITRARY REFERENCE FRAMES

NIKOLAI V. MITSKIEVICH

Nova Science Publishers, Inc.
New York

Copyright © 2006 by Nova Science Publishers, Inc.

All rights reserved. No part of this book may be reproduced, stored in a retrieval system or transmitted in any form or by any means: electronic, electrostatic, magnetic, tape, mechanical photocopying, recording or otherwise without the written permission of the Publisher.

For permission to use material from this book please contact us:
Telephone 631-231-7269; Fax 631-231-8175
Web Site: http://www.novapublishers.com

NOTICE TO THE READER

The Publisher has taken reasonable care in the preparation of this book, but makes no expressed or implied warranty of any kind and assumes no responsibility for any errors or omissions. No liability is assumed for incidental or consequential damages in connection with or arising out of information contained in this book. The Publisher shall not be liable for any special, consequential, or exemplary damages resulting, in whole or in part, from the readers' use of, or reliance upon, this material.

This publication is designed to provide accurate and authoritative information with regard to the subject matter covered herein. It is sold with the clear understanding that the Publisher is not engaged in rendering legal or any other professional services. If legal or any other expert assistance is required, the services of a competent person should be sought. FROM A DECLARATION OF PARTICIPANTS JOINTLY ADOPTED BY A COMMITTEE OF THE AMERICAN BAR ASSOCIATION AND A COMMITTEE OF PUBLISHERS.

LIBRARY OF CONGRESS CATALOGING-IN-PUBLICATION DATA

Relativistic physics in arbitrary reference frames / Nikolai V. Mitskievich, editor.
    p. cm.
Includes bibliographical references and index.
ISBN 1-59454-425-5
1. Relativity (Physics)--Mathematics. 2. Celestial reference systems. 3. Field theory (Physics) I. Mïetìskevich, N. V. (Nikolaæi Vsevolodovich)
QC173.59.M3R45   2005
530.11'01'51--dc22   2005018583

Published by Nova Science Publishers, Inc. ✣ *New York*

Contents

Preface — iii

1 Introduction — 1
1.1 A General Characterization of the Subject 1
1.2 A Synopsis of Notations of Riemannian Geometry 4
1.3 The Noether Theorem: Space-Time Invariance 17
1.4 The Noether Densities' Transformation Laws 28

2 Reference Frames' Calculus — 35
2.1 The Monad Formalism and Its Place in the Description of Reference Frames in Relativistic Physics 35
2.2 Reference Frames Algebra 41
2.3 Geometry of Congruences. Acceleration, Rotation, Expansion and Shear of a Reference Frame 44
2.4 Differential Operations and Identities of the Monad Formalism 47

3 Equations of Motion of Test Particles — 53
3.1 The Electric Field Strength and Magnetic Displacement Vectors 53
3.2 Monad Description of the Motion of a Test Charged Mass in Gravitational and Electromagnetic Fields 56
3.3 Motion of Photons, the Redshift and Doppler Effects 58
3.4 The Dragging Phenomenon 64
    3.4.1 Dragging in Circular Equatorial Orbits in the Kerr Space-Time . . . 66
    3.4.2 An Orbit Shift in the Taub-NUT Space-Time 69
    3.4.3 Dragging in the Space-Time of a Pencil of Light 69
    3.4.4 Other Dragging Effects 74
3.5 More General Gravitoelectromagnetic and Gravitoelectric Phenomena . . . 75

4 The Maxwell Field Equations — 81
4.1 The Four-Dimensional Maxwell Equations 81

	4.2	The Electromagnetic Stress-Energy Tensor and Its Monad Decomposition	82
	4.3	Monad Representation of Maxwell's Equations	85
	4.4	A Charged Fluid Without Electric Field	87
	4.5	An Einstein-Maxwell Field with Kinematic Magnetic Charges	90

5 The Einstein Field Equations — 93

- 5.1 The Four-Dimensional Einstein Equations 93
- 5.2 Monad Representation of Einstein's Equations 94
- 5.3 The Geodesic Deviation Equation and a New Level of Analogy Between Gravitation and Electromagnetism 97
- 5.4 New Quasi-Maxwellian Equations of the Gravitational Field 99
- 5.5 Examples of Space-Times 102
  - 5.5.1 Remarks on Classification of Intrinsic Gravitational Fields 102
  - 5.5.2 Example of the Taub–NUT Field 103
  - 5.5.3 Example of the Spinning Pencil-of-Light Field 106
  - 5.5.4 Gravitational Fields of the Gödel Universe 108

6 Perfect Fluids — 111

- 6.1 Introductive Remarks 111
- 6.2 Rank 2 and 3 Fields 114
- 6.3 Free Rank 2 Field 116
- 6.4 Free Rank 3 Field 117
- 6.5 Rotating Fluids 118
- 6.6 Special Relativistic Theory 120
- 6.7 Additional Remarks 122

7 Canonical Formalism — 123

- 7.1 Mechanics *versus* Field Theory 123
- 7.2 Canonical Approach to Field Theory 127
- 7.3 Canonical Formalism and Quantization 138

8 Concluding Remarks — 147

References — 150

Index — 163

Preface

During many years I have felt a necessity to have at hand a concise, but complete enough, account of the theory of reference frames and its applications to the field theory. There are, of course, such publications as the books by O.S. Ivanitskaya, A.E. Levashyov, V.I. Rodichev, Yu.S. Vladimirov, and by A.L. Zel'manov and V.G. Agakov (all of them in Russian), containing expositions of such ideas, and miscellaneous journal papers. They were however written with the use of notations which differed drastically from the traditional three-vectorial guidelines closely related to intuition-nourishing symbolics, such as time-derivative, spatial divergence and curl, and these presentations have never been sufficiently complete. It was obviously indispensable to translate ideas of the reference frames theory into this more physical language not losing at the same time the rigour of our theory — thus, without restricting to any kind of approximations (which are done quite often in such decent branches of science as the relativistic celestial mechanics where concrete applications are otherwise, of course, unattainable).

The first (much shorter) versions of this book appeared as an article in the collection "The Progress in Science and Technology" (1991, in Russian) and the e-print booklet gr-qc/9606051; the latter received an attention of colleagues working in the reference frames theory who mentioned it in their publications. Now the contents of these earlier versions was thoroughly revised and enlarged, in particular, two new chapters were added which extend application of the monad formalism to qualitatively new parts and branches of field theory: the field theoretic description of perfect fluids and the general covariant canonical formalism.

Acknowledgements

I am greatly indebted to the late Abram L. Zel'manov for a great many enlightening talks on the problems of reference frames in general relativity. He was in fact a great discoverer of new ideas and methods in this field. My thanks are due to Vladimir A. Fock, Anatoly E. Levashyov, Olga S. Ivanitskaya, Alexander S. Kompaneyetz, Vladimir I. Rodichev, Yakov A. Smorodinsky who have helped me so much in different aspects. I am thankful also to Dimitry D. Ivanenko, Alexei Z. Petrov, Yakov P. Terletzky, Evgeny M. Lifshitz, and Alonso Castillo Pérez, of whom my research activities essentially depended at various stages of my scientific career. All of them were very different, I dare say, even controversial people, with really fantastic biographies, and I feel happy for having met them on my personal worldline and having learned so much from their particular "reference frames".

This work could not be accomplished without fruitful collaboration with many of my students; some of them I cannot meet more in this life; many others are my friends and colleagues now. I am afraid that it would be difficult to name here all of them: Viktor N. Zaharov†, José C. del Prado†, Héctor A. Poblete, Maria Ribeiro Teodoro, Alberto A. García, Ildefonso Pulido, Evgeny N. Epikhin, Vladimir E. Markiewicz, Alexander P. Yefremov, David A. Kalev, Salil Gupta, Alexey V. Ul'din, José L. Cindra, Georgios A. Tsalakou, Héctor Vargas Rodríguez, Luis I. Lórez Benítez, and many more, who were not so closely connected with this work on reference frames.

I cordially thank my friends and colleagues for encouragement, real help, brilliant ideas and concrete collaboration: Ernst Schmutzer, Ch. Møller, Bryce S. DeWitt and Cecile Morette DeWitt, Alexander A. Beilinson, Sr., John A. Wheeler, Yuri S. Vladimirov, Valeri D. Zakharov, Kip S. Thorne, Alexander K. Gorbatsievich, Jan Horský, Jerzy F. Plebański, Michael P. Ryan, Jr., Jürgen Ehlers, Vladimir N. Efremov, y otra vez Alberto A. García Díaz.

Rereading these lists I feel that I am taking too much time and patience of the reader, but I do really need to mention with deep gratitude all of these people, as well as to express my sincere acknowledgements of patience and daily support which I have always met in my family (I have to acknowledge very gratefully the generous help extended to me by my stepdaughter Marina in the final preparation of the manuscript of this book). As a matter of fact, I have learned from everybody, and it is my fault not to understand properly many lessons they taught me in my life; thus all the unevenness, inconsistency, or error to be encountered in this book, are indisputably mine.

Chapter 1

Introduction

1.1 A General Characterization of the Subject

The concept of reference frame was introduced in physics at an early stage when its formalization had just begun and even before introduction of systems of coordinates and equations of motion; this concept continues to play a fundamental rôle in our science ever since. When Galileo Galilei first spoke about the principle of relativity of motion [see Schmutzer and Schütz (1983)], he already used in fact the idea of reference frame. Moreover, he closely connected it with his thought experiment method (Gedankenexperiment, later found to be so important also in quantum theory). Still more the concept of reference frames was used by Sir Isaac Newton (1962) when he spoke of the law of inertia (his First Law) and of the nature of time and space; his thought experiment involving a rotating bucket filled with water laid the foundation-stone of Mach's principle. A more concrete description of inertial and non-inertial reference frames was later developed in classical mechanics, together with a study of corresponding effects involving the centrifugal and Coriolis forces [see our discussion of the subject in (Vladimirov, Mitskievich, and Horský 1987)].

In this book, we shall consider mainly the problem of description and further applications of reference frames in relativistic physics primarily using a general relativistic approach to them. The *methods* of general relativity[1] are essential since they inevitably have to be used also in the Minkowski space-time if non-Cartesian coordinates are used. This approach is crucial for interpretation and evaluation of physical effects predicted by the usual four-dimensional theories formulated irrespective to any concrete choice of a reference frame, as well as for determination of physical observables in classical (and, with certain reservations,[2] quantum) theory.

---

[1] We understand the general relativity proper as all relativistic physics taking account of gravitation, *i.e.* of the space-time curvature.

[2] In quantum physics, the measurement procedure implicitly implemented in the very definition of a reference frame in terms of a test reference body, encounters fundamental difficulties due to uncontrollable interaction between the measuring device and quantum object (in fact, the uncertainty relations); see some details in (Bohr and Rosenfeld 1933). See more remarks on p. 40 in this book.

Reference frames are primarily considered in the scope of classical (non-quantum) physics, hence in the assumption that observation and measurement procedures do not influence the behavior of the objects; in particular, we admit that these procedures should not disturb the space-time geometry. This means that such an idealized reference body is a *test* one, while the same physical system of real bodies and fields may be described simultaneously and equally well using any of the infinite set of alternative reference frames. Since the reference body points are test particles, they may be accelerated arbitrarily without application of any finite forces (hence, of any fields), — their masses are infinitesimal too. The only limitation on their motion is due to the relativistic causality principle: since reference body represents an idealization of sets of measuring devices and local observers, its points' world lines should be time-like (any such line lies inside the light cone with its vertex on it). It is not always possible to describe motion of a real system of bodies in a co-moving reference frame without involving singularities (at least, the bodies should not mutually collide), but in order to exclude from our description non-physical singularities, we have to assume that motion of our idealized reference body is described by a *congruence* of time-like world lines.[3] This leads to a fruitful use of ideas of relativistic hydrodynamics in the theory of reference frames; thus a reference body may accelerate, rotate and undergo variable deformation, which completely characterize the reference frame properties. The mathematical description of arbitrary reference frames (including non-inertial ones) and of the corresponding physical effects, should be consistent with the assumed space-time geometry and have an explicitly general covariant form.

Dealing with the reference frame concept sometimes comes across certain psychological obstacles when one either confuses reference frame with system of coordinates (non-covariant approaches), or connects it to any four-dimensional basis (tetrad) whatsoever. Both extremities, in the author's opinion, come together in the sense of gross vulgarization of this merely simple notion; what is much worse, these approaches are widely practised, which involves frequently a disregard and even complete negligence of reference frames in the most parts of physics. We shall revisit this problem in the section 2.1 that precedes the description of the reference frames calculus.

It is remarkable that all algebraic relations in electrodynamics are the same in the standard three-dimensional form and in the formalism of arbitrary reference frames in any gravitational fields. But what is even more important, the (differential) dynamical laws of electromagnetic field theory in non-inertial frames differ from their standard three-dimensional Maxwellian form by additional terms which are naturally interpreted analogously to the inertial forces of the classical mechanics. Moreover, even in the flat Minkowski space-time, non-inertial reference frames involve inevitably and rigorously a curved physical three-space which in the presence of a rotation is also non-holonomic in the sense that there exist only *local* three-space elements which do not form global three-hypersurfaces. Thus the expressions of the inertial forces and their field theoretical generalizations either do not change at all, or change insubstantially when one turns from general to special relativity (only the

---

[3]Congruence is a family of lines such that through any point in the region of its definition passes one and only one line.

four-curvature disappears). For example, Maxwell's equations retain in a non-inertial frame in special relativity exactly the same monad form they have in general relativity. Under a monad, we understand a time-like unit vector field whose integral lines depict worldlines of the "particles" forming a test body of reference (see below). But we have mentioned the rôle of reference frames in interpreting theoretical results and evaluating effects the theory predicts, not in a casual remark: the very solutions of the particles and fields dynamics equations are in general much easier obtained from their standard four-dimensional form using simplifications due to a choice of coordinates which corresponds to the symmetries inherent in the problem under consideration. Only later the use of reference frames comes when it simplifies and stimulates the physical interpretation of these results.

Some remarks about the structure of this book. In section 1.2 we consider mathematical (mainly geometric) definitions and relations. This is necessary since the notations differ substantially from one publication to another, and moreover it is necessary to have at hand some exotic formulae. This paper may also be of use for colleagues working in other areas than gravitation theory, and a bit of self-containedness would help them to get into the subject. The last two sections of chapter 1 are dedicated to the (second) Noether theorem and properties of objects following from it (the Noether densities). These matters are scarcely enough presented in the accessible literature, but are widely used in this book and helpful in dealing with many problems in classical and quantum physics.

In chapter 2, after the already mentioned section 2.1, we give in section 2.2 the basic relations of the monad formalism algebra, including purely spatial but nevertheless generally covariant (in the four-dimensional sense) operations of the scalar and vector products. In further sections of the chapter 2, the differential relations of the monad formalism are given: on the one hand, these are important characteristics related to those of mechanics of continuous media; on the other hand, differential operations as well as the corresponding identities are considered.

Chapters 3 and 4 deal not only with gravitation, but essentially with electrodynamics too. The first of them is dealing with the relativistic mechanics of electrically charged particles, the second with dynamics of the very electromagnetic field. It is worth specially mentioning section 3.2 in which the equations of motion of a test massive charge are considered from the viewpoint of an arbitrary reference frame, 3.3 where examples are given of calculation of the frequency shift effect in three characteristic cases, and 3.4 giving other type of examples, with objectively determined Killing reference frames which enable a 'standard' definition of the dragging phenomenon. The section 3.5 is dedicated to more general cases of gravitoelectromagnetic phenomena which should be of a fundamental importance in physics. Maxwell's equations are written down in an arbitrary reference frame in section 4.3 where non-inertial source terms are calculated and discussed. In section 4.4 the reference frame formalism is applied to physical interpretation of an exact solution of the system of Einstein-Maxwell equations where a non-test charged fluid does not produce any electric field in its co-moving frame. In 4.5 a kinematic magnetic charge density is shown to appear in the space-time of an Einstein-Maxwell *pp*-wave in a quite natural reference frame endowed with rotation. In chapter 5, section 5.2 we consider 3+1-splitting of

Einstein's equations in the monad formalism and trace both analogies and dissimilarities between the electromagnetic and gravitational interactions. Then, in sections 5.3 and 5.4 a new kind of analogy between gravitation and electromagnetism is established in terms of the Weyl conformal curvature tensor. This analogy is practically perfect, and it does not need any *ad hoc* hypotheses concerning the gravitational theory. At the same time, it is shown that the geodesic deviation equation takes a form analogous to that known for motion of a test charge in the Maxwell electrodynamics which involves the usual Lorentz force. In chapter 6 we consider a new approach to description of perfect fluids in terms of fields whose potentials are 2- and 3-forms. The chapter 7 is dedicated to a general covariant monad representation of the canonical formalism in field theory. Finally, in chapter 8 the concluding remarks are made.

1.2 A Synopsis of Notations of Riemannian Geometry

Quite a number of expressions and relations of Riemannian geometry will be extensively used below. In existing literature [see for a traditional approach, *e.g.*, Eisenhart (1926, 1933, 1972); Schouten and Struik (1935, 1938); as to a presentation of general relativity from the mathematical viewpoint, see Sachs and Wu (1977)], they are scattered in different parts of various publications, and the notations differ substantially from one source to another. Many relations are naturally given without derivation. The reader may skip this section but look into it merely for clarification of some formulae. It is worth emphasizing that we take everywhere the speed of light equal to unity ($c = 1$), the space-time signature being $(+---)$; the Greek indices are four-dimensional, running from 0 to 3, the Einstein summation convention is adopted, including collective indices. Symmetrization and antisymmetrization with respect to sets of indices are denoted by the standard *Bach brackets*, $(\cdots)$ and $[\cdots]$ respectively, with the indices in the brackets. Tetrad indices are written in separate parentheses, *e.g.*, $F_{(\mu)(\nu)}$. For important details of abstract representation of the tensor calculus and its applications to field theory, see (Israel 1970, Ryan and Shepley 1975, Eguchi, Gilkey and Hanson 1980, Choquet-Bruhat, DeWitt-Morette and Dillard-Bleick 1982, von Westenholz 1986).

We admit that the tensor algebra in an arbitrary basis and in abstract notations, is commonly known. Hence we begin with the Cartan forms. To make notations shorter, we shall sometimes employ collective indices, *e.g.* in the basis of a *p*-form [*cf.* Mitskievich and Merkulov (1985)]:

$$dx^a := dx^{\alpha_1} \wedge \cdots \wedge dx^{\alpha_p}.$$

Here exterior (wedge) product is supposed to be an antisymmetrization of the tensorial product,

$$dx^{\alpha_1} \wedge \cdots \wedge dx^{\alpha_p} := dx^{[\alpha_1} \otimes \cdots \otimes dx^{\alpha_p]}.$$

Similarly, we shall write tetrad basis of a *p*-form as $\theta^{(a)}$ (then a collective or individual index in the parentheses pertains to a tetrad field); this construction is analogous to dx^a, and represents an exterior product of the covector basis 1-forms, $\theta^{(\alpha)} = g^{(\alpha)}{}_\beta dx^\beta$. Here

$g^{(\alpha)}{}_\beta$ are the covariant tetrad components corresponding to the lower index β of a set of four independent covectors (enumerated by the tetrad index (α)). We write the tetrad index in its individual parentheses immediately after the root letter g, the same as that of the metric tensor, only then followed by the (coordinated) component number (here, β).[4] Then a p-form α can be written as

$$\alpha = \alpha_a dx^a = \alpha_{(a)} \theta^{(a)},$$

while

$$\alpha \wedge \beta = (-1)^{pq} \beta \wedge \alpha, \qquad (1.2.1)$$

where $p = \deg \alpha$, $q = \deg \beta$ (degrees — or ranks — of the forms α and β).

The axial tensor of Levi-Cività is defined with help of the corresponding symbol,[5]

$$E_{\alpha\beta\gamma\delta} = (-g)^{1/2} \epsilon_{\alpha\beta\gamma\delta}, \qquad E^{\alpha\beta\gamma\delta} = -(-g)^{-1/2} \epsilon_{\alpha\beta\gamma\delta}. \qquad (1.2.2)$$

In our four-dimensional (essentially, even-dimensional) manifold, this tensor has the following properties:

$$E_{ag} = (-1)^{p(4-p)} E_{ga} \equiv (-1)^p E_{ga},$$

$$E_{ag} E^{bg} = -p!\,(4-p)!\, \delta_a^b; \qquad \#a = \#b = p = 4 - \#g.$$

While using here collective indices, we denote the number of individual indices contained in them, by $\#$. The Kronecker symbol with collective indices is totally skew by a definition,

$$\delta_a^b \equiv \delta_{\alpha_1\cdots\alpha_p}^{\beta_1\cdots\beta_p} := \delta_{[\alpha_1}^{\beta_1} \cdots \delta_{\alpha_p]}^{\beta_p}, \qquad \delta_{[a}^b \delta_{u]}^v \equiv \delta_{au}^{bv}, \qquad (1.2.3)$$

so that $A^a \delta_a^b \equiv A^b$ (let A^a be skew in all individual indices). Dual conjugation of a form is denoted by the Hodge star $*$ [see Eguchi, Gilkey and Hanson (1980)] before the form. The star acts on the basis,

$$*dx^a := \frac{1}{(4-p)!} E^a{}_g dx^g, \qquad *\alpha := \alpha_a * dx^a,$$

$$*1 = \frac{1}{4!} E_{\alpha\beta\gamma\delta} dx^\alpha \wedge dx^\beta \wedge dx^\gamma \wedge dx^\delta = \sqrt{-g}(dx),$$

---

[4]In the metric space, to which the pseudo-Riemannian space-time belongs, any tensor (in particular, vector) can be taken in its covariant and contravariant form (with respect to any of its indices) due to existence of the metric tensor which in fact is a generalization of a mere identity matrix; thus to the covariant basis $\theta^{(\alpha)}$ always corresponds the contravariant basis $X_{(\alpha)}$: this is the alternative use of the term "dual". However such a duality has nothing in common with the duality involving the Levi-Cività axial tensor.

[5]This means that the Levi-Cività symbol itself can be simultaneously interpreted as a contravariant axial tensor density (of weight $+1$) *and* a covariant axial tensor density of weight -1: $\epsilon_{\alpha\beta\gamma\delta} = -(-g)^{1/2} E^{\alpha\beta\gamma\delta} = (-g)^{-1/2} E_{\alpha\beta\gamma\delta}$. Remember that a tensor density is the corresponding tensor multiplied by $(-g)^{w/2}$, where w is the weight of the tensor density, so that, for example, the transformation law of a scalar density $\mathfrak{L}$ (of weight w; for $w = 1$, as the Lagrangean density usually is denoted), is $\mathfrak{L}'(x') = |J|^{-w} \mathfrak{L}(x)$, $|J|$ being absolute value of the Jacobian of the coordinate transformation. Thus the reader may find some discrepancies on p. 129 of the otherwise quite good book by B.F. Schutz (1980).

(dx) being the four-dimensional volume element corresponding to the four covectors dx^0, dx^1, dx^2, and dx^3. Thus $**\alpha = (-1)^{p+1}\alpha$ (here even-dimensional nature of the manifold is essential); in particular, $**1 = -1$. It is worth stressing that the sequence of the (lower and upper) indices is of great importance too, e.g. it was not the question of a chance when we wrote $E^a{}_g$ and not $E_g{}^a$ [which in fact is equal to $(-1)^p E^a{}_g$, cf. the formula after (1.2.2)].

The usual dual conjugation of a bivector will be also denoted by a star, but this will be written over (or under) the pair of indices to which it is applied, e.g.

$$F^{\kappa\lambda}_{*} := \frac{1}{2} E^{\kappa\lambda\mu\nu} F_{\mu\nu}, \tag{1.2.4}$$

thus $F^{**}_{\alpha\beta} \equiv -F_{\alpha\beta}$. It is obvious that $F_{\kappa\lambda}$ is skew (we called it bivector for this reason), so that the dual conjugation does not lead to any loss of information. Our choice of place for the star is justified here by the convenience in writing down the so-called *crafty identities* where asterisks are applied to different pairs of indices [see e.g. (Mitskievich and Merkulov 1985)]:

$$V^{\kappa\lambda\,*}_{*\;\;\mu\nu} = -V_{\mu\nu}{}^{\kappa\lambda} - \frac{1}{2}\delta^{\kappa\lambda}_{\mu\nu}V_{\sigma\tau}{}^{\sigma\tau} + \delta^{\kappa\tau}_{\mu\nu}V_{\sigma\tau}{}^{\sigma\lambda} + \delta^{\tau\lambda}_{\mu\nu}V_{\sigma\tau}{}^{\sigma\kappa}. \tag{1.2.5}$$

For special cases and when a contraction is performed, one obtains in electrodynamics [cf. (Wheeler 1962, Israel 1970, Mitskievich and Merkulov 1985)]

$$F_{\mu\nu}F^{\lambda\nu} - F^{*}_{\mu\nu}F^{\lambda\nu}_{*} = \frac{1}{2}\delta^\lambda_\mu F_{\sigma\tau}F^{\sigma\tau}, \tag{1.2.6}$$

$$F^{*}_{\mu\nu}F^{\lambda\nu} = \frac{1}{4}\delta^\lambda_\mu F^{*}_{\sigma\tau}F^{\sigma\tau}. \tag{1.2.7}$$

For the Levi-Cività axial tensor one has

$$E^{\kappa\lambda\,*}_{*\;\;\mu\nu} \equiv E^{\kappa\lambda\,*}_{\;\;\mu\nu} = -2\delta^{\kappa\lambda}_{\mu\nu}, \tag{1.2.8}$$

for the Riemann–Christoffel curvature tensor,

$$R^{*\;\kappa\lambda}_{\alpha\beta\;*} = -R_{\alpha\beta}{}^{\kappa\lambda} + R\delta^{\kappa\lambda}_{\alpha\beta} - 4R^{[\kappa}_{[\alpha}\delta^{\lambda]}_{\beta]}, \tag{1.2.9}$$

and for the Weyl conformal curvature tensor,

$$C^{*\;\gamma\delta}_{\alpha\beta} \equiv C_{\alpha\beta}{}^{\gamma\delta\,*}_{\;\;*}. \tag{1.2.10}$$

By the way, a repeated application of the crafty identities to a construction quadratic in the curvature tensor (with subsequent contractions, so that only two indices remain free ones), leads to the well known Lanczos identities,

$$R_{\alpha\beta\gamma}{}^\mu R^{\alpha\beta\gamma\nu} - \tfrac{1}{4}R_{\alpha\beta\gamma\delta}R^{\alpha\beta\gamma\delta}g^{\mu\nu} - 2R^{\mu\alpha\beta\nu}R_{\alpha\beta} + R_{\alpha\beta}R^{\alpha\beta}g^{\mu\nu} -$$
$$- R^\mu_\alpha R^{\alpha\nu} + R\,R^{\mu\nu} - \tfrac{1}{4}R^2 g^{\mu\nu} = 0 \tag{1.2.11}$$

[see our derivation of these identities (Mitskievich 1969) which differs radically from that by Lanczos (1938) who used less straightforward integral relations]. Note that the invariant $R_{\alpha\beta\gamma\delta}R^{\alpha\beta\ \gamma\delta}_{*\ \ *}$ is known as the Bach–Lanczos invariant, or the generalized Gauss–Bonnet invariant;[6] like the electromagnetic invariant $F_{\alpha\beta}F^{\alpha\beta}_{\ \ *}$, it represents (in a four-dimensional world) a pure divergence, so that it does not contribute to the field equations in four dimensions. *Cf.* also DeWitt (1965).

Scalar multiplication of Cartan's forms is realized with help of consecutive dual conjugations:

$$*(dx_k \wedge *dx^{al}) = (-1)^{p+1}\frac{(p+q)!}{p!}\delta^{[l}_{\ k}dx^{a]}, \qquad \begin{cases} \#a = p, \\ \#k = q = \#l \end{cases}$$

in particular,

$$*(dx^\lambda \wedge *dx^\mu) = -g^{\lambda\mu}.$$

The last relation is equivalent to

$$*(E^\mu_{\ \pi\rho\sigma}dx^{\lambda\pi\rho\sigma}) = E^\mu_{\ \pi\rho\sigma}E^{\lambda\pi\rho\sigma} = -3!\, g^{\lambda\mu},$$

as well as to

$$dx^\alpha \cdot dx^\beta = g^{\alpha\beta}$$

(in fact, this property should be considered as introduction of a symmetric metric tensor g in the manifold under consideration). For the more general tetrad basis θ^α we have

$$\theta^\alpha \cdot \theta^\beta = g^{(\alpha)(\beta)}.$$

Contravariant objects are treated with the use of the corresponding coordinated vector basis ∂_μ or tetrad vector basis $X_{(\mu)}$ (where necessary, their tensor products). Here ∂_μ is usually considered as the partial differentiation operator $\frac{\partial}{\partial x^\mu}$ acting on (scalar) functions; see however comments in the next footnotes. The scalar product operation is defined so that $dx^\alpha \cdot \partial_\mu \equiv \partial_\mu \cdot dx^\alpha = \delta^\alpha_\mu$. The tetrad basis is related to its tetrad components (the covariant case was mentioned above; in the contravariant one, $X_{(\sigma)} = g_{(\sigma)}^{\ \mu}\partial_\mu$, hence $X_{(\sigma)} \cdot dx^\alpha = g_{(\sigma)}^{\ \alpha}$, while $\theta^{(\sigma)} \cdot \partial_\alpha = g^{(\sigma)}_{\ \alpha}$). Thus there always exists a pair of "conjugated" tetrads: one is the set of linearly independent contravariant vectors, $X_{(\sigma)}$, and another, a set of linearly independent covariant vectors, $\theta^{(\sigma)}$, so that $g_{(\sigma)}^{\ \alpha}g^{(\sigma)}_{\ \beta} = \delta^\alpha_\beta$ and, equivalently, $g_{(\rho)}^{\ \alpha}g^{(\sigma)}_{\ \alpha} = \delta^\sigma_\rho$: remember the alternative concept of "duality" mentioned above (not to be confused with the Hodge conjugation).

The covariant differentiation axioms [see *e.g.* Ryan and Shepley (1975)] are:
(1) $\nabla_v T$ has the same tensor rank and valence properties as T, v being a vector.
(2) ∇_v is a linear operation:

$$\nabla_v(T + U) = \nabla_v T + \nabla_v U,$$

---
[6]The first of Lagrangians of the Lovelock (1971) series reduces to it.

$$\nabla_{au+bv}T = a\nabla_u T + b\nabla_v T.$$

(3) The Leibniz property[7] holds (including the contraction operation in the Riemannian geometry).
(4) Action of ∇ on a function[8] is $\nabla_v f = vf$, where $v = v^\alpha \partial_\alpha \equiv v^{(\alpha)} X_{(\alpha)}$.
(5) Metricity property: $\nabla_v g = 0$ (for a further generalization of the geometry, the non-metricity tensor appears on the right-hand side).
(6) Zero torsion axiom:

$$\nabla_u v - \nabla_v u = [u, v] + \mathbb{T}, \quad \mathbb{T} = 0$$

(in a generalization of the geometry, the torsion tensor[9] $\mathbb{T}$ becomes different from zero).

These axioms[10] are most simply realized when the connection coefficients are introduced,

$$\nabla_{X_{(\alpha)}} X_{(\beta)} = \Gamma^{(\gamma)}_{(\beta)(\alpha)} X_{(\gamma)}; \quad \nabla_{X_{(\alpha)}} \theta^{(\gamma)} = -\Gamma^{(\gamma)}_{(\beta)(\alpha)} \theta^{(\beta)} \qquad (1.2.12)$$

(only one of these relations is independent). When the structure coefficients of a basis are introduced with help of

$$[X_{(\alpha)}, X_{(\beta)}] = C^{(\gamma)}_{(\alpha)(\beta)} X_{(\gamma)}, \quad C^{(\gamma)}_{(\alpha)(\beta)} = \Gamma^{(\gamma)}_{(\beta)(\alpha)} - \Gamma^{(\gamma)}_{(\alpha)(\beta)} \qquad (1.2.13)$$

(for a holonomic, or coordinated basis, $C^\gamma_{\alpha\beta} = 0$; as usually, we write in individual parentheses only the tetrad and not coordinated basis indices), the general solution satisfying the whole set of axioms, from 1 to 6, is

$$\Gamma^{(\gamma)}_{(\alpha)(\beta)} = \frac{1}{2} g^{(\gamma)(\delta)} \left[X_{(\beta)} g_{(\alpha)(\delta)} + X_{(\alpha)} g_{(\beta)(\delta)} - X_{(\delta)} g_{(\alpha)(\beta)} \right]$$
$$+ \frac{1}{2} \left[C^{(\gamma)}_{(\beta)(\alpha)} + C^{(\delta)}_{(\epsilon)(\beta)} g_{(\alpha)(\delta)} g^{(\epsilon)(\gamma)} + C^{(\delta)}_{(\epsilon)(\alpha)} g_{(\beta)(\delta)} g^{(\epsilon)(\gamma)} \right]. \qquad (1.2.14)$$

In a coordinated basis (where the basis vectors and covectors are usually written as ∂_α and dx^α), the connection coefficients reduce to the Christoffel symbols (which are written

---

[7] Distributive property when product expressions are differentiated.

[8] A "scalar"; the concept of scalar is somewhat different for the covariant differentiation denoted by a semicolon; see more in the footnote 11.

[9] Like the curvature operator below, the torsion $\mathbb{T}$ (if $\neq 0$) here also describes a tensor, now of the rank three, by the rule: $\nabla_{X_{(\mu)}} X_{(\nu)} - \nabla_{X_{(\nu)}} X_{(\mu)} - [X_{(\mu)}, X_{(\nu)}] = T^{(\lambda)}_{(\mu)(\nu)} X_{(\lambda)}$ [cf. (1.2.26)]. Then, of course, the connection coefficients following from (1.2.12), should be modified.

[10] There could be also added the seventh (in some sense) "axiom": $[\nabla_u, \nabla_v] = \nabla_{[u,v]} + \mathbb{R}(u, v)$ [cf. the definition (1.2.25)], with $\mathbb{R} = 0$, as an alternative to $\mathbb{T} = 0$. Then, of course, the theory changes drastically, but in literature one can find an assertion that in this case the former gravitational effects become those of the torsion. The author is not ready to swear that this is completely true, and the following exposition strictly belongs to the Riemannian geometry.

without parentheses before and after each index, and which obviously are symmetric in the two lower indices):

$$\Gamma^{\gamma}_{\alpha\beta} = \frac{1}{2}g^{\gamma\delta}\left(g_{\alpha\delta,\beta} + g_{\delta\beta,\alpha} - g_{\alpha\beta,\delta}\right). \quad (1.2.15)$$

Orthonormal bases and those of Newman–Penrose [see Penrose and Rindler (1984a, 1984b)] are special cases of bases whose vectors have constant scalar products (*i.e.*, the corresponding tetrad components of the metric are constant). For such bases the connection coefficients are skew in the upper and the first lower indices; they are sometimes called Ricci rotation coefficients. Thus, in a coordinated basis, the differentiation of $A = A_\alpha dx^\alpha$ yields

$$\nabla_{\partial_\gamma} A = (\partial_\gamma A_\alpha)dx^\alpha + A_\alpha \nabla_{\partial_\gamma} dx^\alpha = \left(A_{\alpha,\gamma} - A_\beta \Gamma^\beta_{\gamma\alpha}\right)dx^\alpha =: A_{\alpha;\gamma}dx^\alpha,$$

and for $V = V^\alpha \partial_\alpha$,

$$\nabla_{\partial_\gamma} V = \left(V^\alpha{}_{,\gamma} + V^\beta \Gamma^\alpha_{\gamma\beta}\right)\partial_\alpha =: V^\alpha{}_{;\gamma} \partial_\alpha,$$

in accordance with the (old) traditional notations for the ;-covariant derivative. From these definitions it is easy to see that the ;-covariant derivatives of vectors, $A_{\alpha;\gamma}$ and $V^\alpha{}_{;\gamma}$, are components of rank 2 tensors. In general the ;-covariant derivatives of rank r tensors also are (in the sense of components) tensors of the rank $r+1$ (*cf.* the axiom 1 of the ∇-covariant derivative).

Let us revisit the definition of connection coefficients, using now the coordinated basis. Then $dx^\alpha = \delta^\alpha_\beta dx^\beta$; it is probably better to highlight more the sense of the index α in this expression (that it is not a component number of a vector, but the number of a covector in a coordinated basis), entering it into a circle (the parentheses already have other meanings): $dx^{@} = \delta^{@}_\beta dx^\beta$. Then

$$\delta^{@}_{\beta;\gamma} = \delta^{@}_{\beta,\gamma} - \Gamma^\sigma_{\gamma\beta}\delta^{@}_\sigma = -\Gamma^{@}_{\gamma\beta},$$

since the Kronecker delta is a set of constants, and the encircled index does not pertain to the coordinated components (now it has more in common with a tetrad component number, but for a non-orthonormal tetrad). Thus we come to an equivalent of (1.2.12) in a coordinated basis, $\nabla_{\partial_\gamma} dx^\alpha = -\Gamma^\alpha_{\gamma\beta} dx^\beta$. The subindex in $\partial_{@} = \delta^\beta_{@} \partial_\beta$ can be treated similarly.

Thus, covariant differentiation can be denoted by a semicolon (;) before the differentiation index (especially, in the elder literature[11]). Then, according to the covariant differentiation axioms (in particular, the axiom 1, however paradoxical this may appear), it leads to an increase by unity of the valence of the corresponding tensor. As Trautman (1956, 1957) remarked, such a covariant derivative can be defined as

$$T_{a;\alpha} := T_{a,\alpha} + T_a|^\tau_\sigma \Gamma^\sigma_{\tau\alpha}, \quad (1.2.16)$$

---

[11] In the old-fashioned coordinated component notations, tensor components are usually treated as synonyms of the corresponding tensors, so it is then quite awkward to consider them as a set of scalars, but in the modern abstract (coordinate-free) notations, tensor — and connection — components are dealt with just as a set of scalar functions. In this book we use a fairly innocuous mixture of both styles.

where the coefficients $T_a|_\sigma^\tau$ determine the behavior of the quantity T_a (which could be not merely a tensor, but also a tensor density or some more general object) under infinitesimal transformations of coordinates, ϵ being a dimensionless infinitesimal constant:

$$x'^\mu = x^\mu + \epsilon \xi^\mu(x),$$

$$\delta T_a := T'_a(x') - T_a(x) = \epsilon T_a|_\sigma^\tau \xi^\sigma{}_{,\tau}, \qquad (1.2.17)$$

i.e. $T_a|_\sigma^\tau = \epsilon^{-1} \partial \delta T_a / \partial(\xi^\sigma{}_{,\tau})$. It is clear that the coefficients $T_a|_\sigma^\tau$ possess a property

$$(S_a T_b)|_\sigma^\tau = S_a|_\sigma^\tau T_b + S_a T_b|_\sigma^\tau, \qquad (1.2.18)$$

which is closely related to the Leibniz property of differentiation. The covariant (in the sense of ∇_u) differentiation axioms are readily reformulated in terms of ;-differentiation; in particular, the axiom 5 takes the form

$$g_{\alpha\beta;\gamma} = 0 \quad \text{or, equivalently,} \quad E_{\kappa\lambda\mu\nu;\gamma} = 0 \qquad (1.2.19)$$

which (in each of these two forms) immediately yields the definition (1.2.15) of $\Gamma^\gamma_{\alpha\beta}$.

The definition of covariant derivative (1.2.16) may be deduced from the concept of Lie derivative,

$$\pounds_\xi T_a := T_{a,\alpha}\xi^\alpha - T_a|_\sigma^\tau \xi^\sigma{}_{,\tau} \equiv T_{a;\alpha}\xi^\alpha - T_a|_\sigma^\tau \xi^\sigma{}_{;\tau}, \qquad (1.2.20)$$

which can be written for objects of the most general nature as

$$\pounds_\xi T_a := \epsilon^{-1}\left(T_a(x) - T'_a(x)\right) \qquad (1.2.21)$$

[cf. Yano (1955)].[12] Due to the property (1.2.18), the definition (1.2.20) of the Lie derivative under an infinitesimal transformation yields the Leibniz property.

However, if we consider the finite-difference 'Lie derivative' (1.2.21) when it is not subjected to an infinitesimal transport (the parameter ϵ is now finite, so that we have to introduce instead of $\pounds$ a new notation, say, $\overset{\bowtie}{\pounds}$), then a generalized Leibniz property has to be introduced,

$$\overset{\bowtie}{\pounds}_{\epsilon\xi}(S_a T_b) = -\left(\overset{\bowtie}{\pounds}_{\epsilon\xi} S_a\right)\left(\overset{\bowtie}{\pounds}_{\epsilon\xi} T_b\right) + \left(\overset{\bowtie}{\pounds}_{\epsilon\xi} S_a\right) T_b + S_a\left(\overset{\bowtie}{\pounds}_{\epsilon\xi} T_b\right).$$

---

[12]The quantity $T'_a(x)$ which will appear also in the proof of Noether's theorem, describes the change of dependence of the components T_a due to the transformation of coordinates from x^μ to x'^μ. This may seem to be somewhat artificial, so we give here an explanation for it in a form of parable. Consider a region of space-time filled with integral curves of the vector field $\xi^\mu(x)$ and two points, P and P', belonging to the same curve, such that the coordinates of P' in the system $\{x'\}$ are numerically equal to those of P in $\{x\}$: $x'^\mu|_{P'} = x^\mu|_P$. This corresponds to the situation when at both points there are independent observers having exactly the same mentality so that they ascribe to their respective positions in space-time the same numerical characteristics thus using these two coordinate systems, $\{x\}$ and $\{x'\}$, respectively. Let these observers have all necessary devices for measurement and reproduction of the field T_a at their respective points. Moreover, suppose that each of them can communicate with the other telepathically. Then the information about measurement of T_a made by the observer at P', sent instantaneously to the observer at P and reproduced by him, is related to the independent measurement of T_a by this latter observer directly through the Lie derivative (1.2.21).

Seemingly, it is quite different from the former one (see the first right-hand side term). However this situation is typical for finite differences, automatically changing to the usual Leibniz property when one goes to the infinitesimal case (in the lower order of ϵ), and only the last two right-hand side terms survive.

It is important that both objects in the parentheses in (1.2.21), have the same argument (non-primed one). This leads to a possibility to extend the operation $T_{a;\alpha}$ even to quantities of the type of a connection. Without going into details — see however (1.3.6) — we would note that in this sense

$$\Gamma^{\lambda}_{\mu\nu;\alpha} = R^{\lambda}{}_{(\mu\nu)\alpha}.$$

The Lie derivative (1.2.20) reduces to the gradient projected on ξ when it is applied to a scalar function. Since the set of coordinates x^{μ} is a collection of four scalar functions, this means that

$$\pounds_{\xi} x^{\mu} = \xi^{\mu}$$

for any vector field ξ.

Concerning the Lie derivative, it is worth also mentioning the relations

$$\pounds_{\xi} g_{\mu\nu} = \xi_{\mu;\nu} + \xi_{\nu;\mu}, \tag{1.2.22}$$

$$\pounds_{\xi} \Gamma^{\lambda}_{\mu\nu} = \xi^{\lambda}{}_{;\mu;\nu} - \xi^{\kappa} R^{\lambda}{}_{\mu\nu\kappa} \equiv \xi^{\lambda}{}_{(;\mu;\nu)} - \xi^{\kappa} R^{\lambda}{}_{(\mu\nu)\kappa}. \tag{1.2.23}$$

A vector field satisfying the equation

$$\xi_{\mu;\nu} + \xi_{\nu;\mu} = 0, \tag{1.2.24}$$

is called the Killing field. Its existence imposes certain limitations on the geometry, — existence of an *isometry*: $\pounds_{\xi} g_{\mu\nu} = 0$ [see (Yano 1955, Eisenhart 1933, Ryan and Shepley 1975)]. The Lie derivative plays an important role in the monad formalism, as we shall see below; this is why a version of this formalism well adapted to description of the canonical structure of the field theory, is called the "Lie-monad formalism" [see (Mitskievich and Nesterov 1981, Mitskievich, Yefremov and Nesterov 1985)].

It is convenient to define the concept of curvature in the Riemannian geometry by introducing the curvature operator [see *e.g.* (Ryan and Shepley 1975)],

$$\mathbb{R}(u,v) := \nabla_u \nabla_v - \nabla_v \nabla_u - \nabla_{[u,v]}. \tag{1.2.25}$$

This operator has the following five properties:
1) It is skew in u and v.
2) It annihilates the metric, $\mathbb{R}(u,v)\, g = 0$.
3) It annihilates any scalar function, $\mathbb{R}(u,v)\, f = 0$ (e.g., $\mathbb{R}(u,v)\, \Gamma^{\alpha}_{\beta\gamma} = 0$).
4) This is a linear operator (*cf.* the axiom 2 of the covariant differentiation).
5) The Leibniz property holds for it (this fact is remarkable since the curvature operator involves the second, not first order differentiation).

The curvature tensor components (which are scalars from the point of view of the ∇-differentiation) are determined as

$$R^{(\kappa)}{}_{(\lambda)(\mu)(\nu)} := \theta^{(\kappa)}\mathbb{R}(X_{(\mu)}, X_{(\nu)})X_{(\lambda)}. \tag{1.2.26}$$

These are the coefficients (components) of decomposition of the curvature tensor with respect to the corresponding basis, and in the general case they are

$$R^{(\alpha)}{}_{(\beta)(\gamma)(\delta)} = X_{(\gamma)}\Gamma^{(\alpha)}_{(\beta)(\delta)} - X_{(\delta)}\Gamma^{(\alpha)}_{(\beta)(\gamma)} + \Gamma^{(\epsilon)}_{(\beta)(\delta)}\Gamma^{(\alpha)}_{(\epsilon)(\gamma)}$$
$$- \Gamma^{(\epsilon)}_{(\beta)(\gamma)}\Gamma^{(\alpha)}_{(\epsilon)(\delta)} - C^{(\epsilon)}_{(\gamma)(\delta)}\Gamma^{(\alpha)}_{(\beta)(\epsilon)}. \tag{1.2.27}$$

It is worth remembering that some authors use a definition of curvature with the opposite sign, *e.g.* Penrose and Rindler (1984a,b) and Yano (1955) [see also a table in (Misner, Thorne and Wheeler 1973)]. The structure of the curvature operator yields a simple property,

$$R^{\kappa}{}_{\lambda\mu\nu}w^{\lambda} = w^{\kappa}{}_{;\nu;\mu} - w^{\kappa}{}_{;\mu;\nu}. \tag{1.2.28}$$

It is also easy to show that the standard algebraic identities hold,

$$R_{\kappa\lambda\mu\nu} = R_{[\kappa\lambda][\mu\nu]} = R_{\mu\nu\kappa\lambda},$$
$$R_{\kappa[\lambda\mu\nu]} = 0, \text{ or equivalently, } R^{*\nu\lambda}{}_{\mu\nu} = 0 \tag{1.2.29}$$

(this last identity bears the name of Ricci, but sometimes it is called "algebraic Bianchi identity"). Important role is also played by the differential Bianchi identity,

$$R_{\kappa\lambda[\mu\nu;\xi]} = 0, \text{ equivalently, } R^{*\alpha\beta}{}_{\kappa\lambda;\beta} = 0. \tag{1.2.30}$$

The Ricci curvature tensor we define as $R_{\mu\nu} \equiv R_{\nu\mu} := R^{\lambda}{}_{\mu\nu\lambda}$, *cf.* Eisenhart (1926, 1933), Landau and Lifshitz (1971), and Misner, Thorne and Wheeler (1973). There are also introduced the scalar curvature, $R := R^{\alpha}_{\alpha}$, the Einstein tensor $G_{\mu\nu} = R_{\mu\nu} - (1/2)g_{\mu\nu}R$ forming the left-hand side of Einstein's equations, and the Weyl conformal curvature tensor C,

$$C_{\kappa\lambda\mu\nu} := R_{\kappa\lambda\mu\nu} + g_{\kappa[\mu}R_{\nu]\lambda} + g_{\lambda[\nu}R_{\mu]\kappa} - \frac{1}{3}Rg_{\kappa[\mu}g_{\nu]\lambda}. \tag{1.2.31}$$

The latter (taken with one contravariant index, and in a coordinated basis), does not feel a multiplication of the metric tensor by an arbitrary good function, so it is *conformally invariant*. Remember that the Weyl tensor works only for space-time dimensionality $D \geq 4$. When $D = 2$, $G_{\mu\nu} \equiv 0$, and for $D = 3$ the Cotton tensor (Cotton 1899) is used when the conformal correspondence of space-times and the classification à la Petrov are considered, see (Garcia *et al.* 2004).

In the formalism of Cartan's exterior differential forms, see (Israel 1970, Ryan and Shepley 1975), the operation of exterior differentiation is introduced,

$$d := \theta^{(\alpha)}\nabla_{X_{(\alpha)}}\wedge \equiv dx^{\alpha}\nabla_{\partial_{\alpha}}\wedge, \tag{1.2.32}$$

where $\wedge$ is the discussed above wedge product (the operation of exterior multiplication), so that $dd \equiv 0$. The operation d has its most simple representation in a coordinated basis (since the Christoffel symbols are symmetric in their two lower indices). It is however especially convenient to work in an orthonormalized or Newman–Penrose basis when the connection 1-forms

$$\omega^{(\alpha)}{}_{(\beta)} := \Gamma^{(\alpha)}{}_{(\beta)(\gamma)} \theta^{(\gamma)}$$

are skew-symmetric ($\omega^{(\alpha)}{}_{(\beta)} = -\omega_{(\beta)}{}^{(\alpha)}$). In the general case they are found as a solution of the system of *Cartan's first structure equations*,

$$d\theta^{(\alpha)} = -\omega^{(\alpha)}{}_{(\beta)} \wedge \theta^{(\beta)}, \tag{1.2.33}$$

and relations for differential of the tetrad metric components,

$$dg_{(\alpha)(\beta)} = \omega_{(\alpha)(\beta)} + \omega_{(\beta)(\alpha)}.$$

Then one has just 24 + 40 equations for determining all the 64 components of the connection coefficients, so that the solution of this set of equations should be unique. It is easy to accustom to find the specific solutions without performing tedious formal calculations; then the work takes surprisingly short time. Since reference frame characteristics (vectors of acceleration and rotation, and rate-of-strain tensor) below will turn out to be a kind of connection coefficients, this approach allows to calculate promptly also these physical characteristics, though for the rate-of-strain tensor the computation will be rather specific.

In their turn, the curvature tensor components are also computed easily and promptly, if one applies *Cartan's second structure equations*,

$$\Omega^{(\alpha)}{}_{(\beta)} = d\omega^{(\alpha)}{}_{(\beta)} + \omega^{(\alpha)}{}_{(\gamma)} \wedge \omega^{(\gamma)}{}_{(\beta)}. \tag{1.2.34}$$

Here the calculations are much more straightforward than for the connection 1-forms, and they take less time. It remains only to read out expressions for the curvature components according to the definition

$$\Omega^{(\alpha)}{}_{(\beta)} := \frac{1}{2} R^{(\alpha)}{}_{(\beta)(\gamma)(\delta)} \theta^{(\gamma)} \wedge \theta^{(\delta)}$$

being in conformity with the curvature operator properties *via* (1.2.34). The Ricci and Bianchi identities read in the language of Cartan's forms as

$$\Omega^{(\alpha)}{}_{(\beta)} \wedge \theta^{(\beta)} = 0 \tag{1.2.35}$$

and

$$d\Omega^{(\alpha)}{}_{(\beta)} = \Omega^{(\alpha)}{}_{(\gamma)} \wedge \omega^{(\gamma)}{}_{(\beta)} - \omega^{(\alpha)}{}_{(\gamma)} \wedge \Omega^{(\gamma)}{}_{(\beta)} \tag{1.2.36}$$

respectively.

Let us give here a scheme of definition of transports. Denote first as $\delta_T v := v|_Q - v|_{P \to Q}$ the change of a vector (or of some other object) under its transport between points P and Q [see (Mitskievich, Yefremov and Nesterov 1985)]. Then for the parallel transport

$$\frac{\delta_P v}{d\lambda} = \nabla_u v \qquad (1.2.37)$$

and for the Fermi–Walker transport [*cf.* Synge (1960)]

$$\frac{\delta_{FW} v}{d\lambda} = \nabla_u v + \eta[(v \cdot \nabla_u u)u - (v \cdot u)\nabla_u u] =: \nabla_u^{FW} v, \qquad (1.2.38)$$

where $u^\alpha = dx^\alpha/d\lambda$ is the tangent vector to the transport curve (for the Fermi–Walker transport, it is essential to choose canonical parameter along the transport curve in such a way that the norm $u \cdot u = \eta^{-1}$ would be constant and non-zero — so the curve itself should be non-null for the Fermi–Walker transport). Using the definitions (1.2.16), it is easy to generalize the transport relations to arbitrary tensors or tensor densities; so for the Fermi–Walker transport

$$\frac{\delta_{FW} T_a}{d\lambda} = T_{a;\alpha} u^\alpha + \eta T_a|_\sigma^\tau (u_{\tau;\alpha} u^\sigma - u_\tau u^\sigma{}_{;\alpha}) u^\alpha. \qquad (1.2.39)$$

From (1.2.38), a definition of the Fermi–Walker connection seems to be directly obtainable, together with the corresponding covariant FW differentiation operation which should lead also to the FW-curvature [*cf.* (Mitskievich, Yefremov and Nesterov 1985)]:

$$\mathbb{R}^{FW}(w, v) := \nabla_u^{FW} \nabla_v^{FW} - \nabla_v^{FW} \nabla_u^{FW} - \nabla_{[u,v]}^{FW}. \qquad (1.2.40)$$

The reader may find that in this generalization a difficulty arises: the "derivative" ∇_u^{FW} does not possess the linearity property. This trouble is however readily removable if we begin with introduction of another type of derivative, namely ∇_u^τ (see below) which possesses all necessary properties of a covariant derivative.

We give here the (equivalent to each other) forms of the geodesic equation (describing parallel transport of a vector along itself) for a canonical (affine) parameter (then absolute value of the vector does not suffer changes):

$$\left. \begin{aligned} & \nabla_u u = 0, \\[4pt] & u \wedge *du = 0, \\[4pt] & \frac{d^2 x^\alpha}{d\lambda^2} + \Gamma^\alpha_{\beta\gamma} \frac{dx^\beta}{d\lambda} \frac{dx^\gamma}{d\lambda} = 0, \\[4pt] & \frac{d}{d\lambda}\left(g_{\alpha\beta} \frac{dx^\beta}{d\lambda}\right) = \frac{1}{2} g_{\sigma\tau,\alpha} \frac{dx^\sigma}{d\lambda} \frac{dx^\tau}{d\lambda} \end{aligned} \right\}. \qquad (1.2.41)$$

A geodesic may be, of course, also null (lightlike), but if it is time-like, one has $d\lambda = ds$. If a non-canonical parameter is used, an extra term proportional to u should be added in the equations (1.2.41), with a coefficient being some function of the (non-canonical) parameter.

Below we shall make use of the just mentioned generalization of the FW-derivative (1.2.38) along a time-like congruence with a tangent vector field τ normalized by unity (the *monad*). It is convenient to denote such a derivative as ∇_u^τ,

$$\nabla_u^\tau T = \nabla_u T + T_a|_\nu^\mu \left(\tau_{\mu;\alpha}\tau^\nu - \tau_\mu \tau^\nu{}_{;\alpha}\right) u^\alpha B^a, \qquad (1.2.42)$$

where B^a is the (coordinated) basis of T: $T = T_a B^a$. Then $\nabla_u^{\text{FW}} \equiv \nabla_u^u$. The definition (1.2.42) can be also written as

$$T_{a;\mu}^{[\tau]} = T_{a;\mu} + T_a|_\sigma^\rho (\tau^\sigma \tau_{\rho;\mu} - \tau^\sigma{}_{;\mu}\tau_\rho). \qquad (1.2.43)$$

This τ-derivative acts as the usual ∇_u upon any scalar function f, and it annuls the vector τ, the metric tensor g, and the construction $b = g - \tau \otimes \tau$ (the three-metric or the projector in the monad formalism):

$$\nabla_u^\tau f = uf, \qquad \nabla_u^\tau \tau = \nabla_u^\tau g = \nabla_u^\tau b = 0 \qquad (1.2.44)$$

(action of a τ-derivative on a tensor is understood in the sense of (1.2.42); u is considered as a partial differentiation operator in uf). The corresponding τ-curvature operator is denoted as $\mathbb{R}^\tau(w,v)$ [see (2.4.10) and (Mitskievich, Yefremov and Nesterov 1985)].

Alongside with exterior differentiation d which increases the degree of a form by unity, in Cartan's formalism another differential operation is also used which diminishes this degree by unity,

$$\delta := -*d*. \qquad (1.2.45)$$

In fact this is the divergence operation,

$$\delta(\alpha_{\mu h} dx^{\mu h}) = p\alpha_{\mu h}{}^{;\mu} dx^h, \qquad p = \#h + 1, \qquad (1.2.46)$$

which does not however fulfil the Leibniz property: when applied to a product, *e.g.*, of a function and a 1-form, fa, it results in

$$\delta(fa) = f\delta a - *(df \wedge *a),$$

while it acts on a function (0-form) simply annihilating it, $\delta f \equiv 0$. One might call this a generalized Leibniz property. Of course, similarly to the identity $dd \equiv 0$, there holds also the identity $\delta\delta \equiv 0$. A combination of both operations gives the de Rham operator (de-Rhamian),[13]

$$\triangle := d\delta + \delta d \equiv (d + \delta)^2 \qquad (1.2.47)$$

(remember divergence of a gradient in Euclidean geometry). A form annihilated by the de-Rhamian, is called harmonic form (remember harmonic functions and the Laplace operator). In contrast to δ, the de-Rhamian acts non-trivially on 0-forms, and in contrast to

---

[13] The same symbol, $\triangle$, also denotes the usual Laplacian.

the exterior differential d, it acts non-trivially on 4-forms too (also in the four-dimensional world). The following classification elements of forms are also used: an exact form is that which is an exterior derivative of another form; a closed form is that whose exterior differential vanishes (an exact form is always closed, though the converse is in general not true); a co-exact form is a result of action by δ on some form of a higher (by unity) degree; a co-closed form is itself identically annihilated by δ. In (topologically) simplest cases it is possible to split an arbitrary form into sum of a harmonic, exact, and co-exact forms [see Eguchi, Gilkey and Hanson (1980), Mitskievich, Yefremov and Nesterov (1985)].

For an arbitrary form α,

$$\triangle \alpha = \alpha_{a;\mu}{}^{;\mu} dx^a - p R^\sigma{}_\tau{}^\mu{}_\nu \alpha_{\mu h}|^\tau_\sigma dx^{\nu h}, \quad \#a = p. \tag{1.2.48}$$

The number of individual indices entering the collective index a of the components of the form α, may be any from 0 to 4, while the quantity $\alpha_a|^\tau_\sigma$ is determined by (1.2.17). The expression (1.2.48) can be reduced identically to

$$\triangle \alpha = (p!)^{-1} [\alpha_{\mu\nu h;\lambda}{}^{;\lambda} + p R^\sigma_\mu \alpha_{\sigma \nu h} + \frac{p(p-1)}{2} R^{\sigma\tau}{}_{\mu\nu} \alpha_{\sigma\tau h}] dx^{\mu\nu h}, \tag{1.2.49}$$

where $a = \mu\nu h$, $\#a = 2 + \#h = p$, and p may be equal also to zero.

We compare here explicitly the results of action of de-Rhamiam on forms of all degrees in a four-dimensional world:

$$f: \quad \triangle f = f_{,\mu}{}^{;\mu}, \tag{1.2.50}$$

$$A = A_\alpha dx^\alpha: \quad \triangle A = (A_{\alpha;\mu}{}^{;\mu} + R^\sigma_\alpha A_\sigma) dx^\alpha, \tag{1.2.51}$$

$$F = \frac{1}{2} F_{\alpha\beta} dx^{\alpha\beta}: \quad \triangle F = \frac{1}{2} (F_{\alpha\beta;\mu}{}^{;\mu} + 2 R^\sigma_\alpha F_{\sigma\beta} + R^{\sigma\tau}{}_{\alpha\beta} F_{\sigma\tau}) dx^{\alpha\beta}, \tag{1.2.52}$$

$$T = \frac{1}{3!} T_{\alpha\beta\gamma} dx^{\alpha\beta\gamma}: \quad \triangle T = \frac{1}{3!} (T_{\alpha\beta\gamma;\mu}{}^{;\mu} + 3 R^\sigma_\alpha T_{\sigma\beta\gamma} + 3 R^{\sigma\tau}{}_{\alpha\beta} T_{\sigma\tau\gamma}) dx^{\alpha\beta\gamma}$$

$$= (\tilde{A}_{\alpha;\mu}^{;\mu} + R^\sigma_\alpha \tilde{A}_\sigma) * dx^\alpha, \tag{1.2.53}$$

$$V = \frac{1}{4!} V_{\alpha\beta\gamma\delta} dx^{\alpha\beta\gamma\delta}: \quad \triangle V = \frac{1}{4!} V_{\alpha\beta\gamma\delta;\mu}{}^{;\mu} dx^{\alpha\beta\gamma\delta} = \tilde{f}_{;\mu}^{;\mu} * 1. \tag{1.2.54}$$

From the last two expressions it is clear that there exists a symmetry with respect to an exchange of a basis form to its dual conjugate with a simultaneous dual conjugation of components of the form [compare (1.2.50) and (1.2.54), (1.2.51) and (1.2.53)]. Here we made use of obvious definitions

$$\tilde{A}^\delta = -\frac{1}{3!} T_{\alpha\beta\gamma} E^{\alpha\beta\gamma\delta}, \quad \tilde{f} = -\frac{1}{4!} V_{\alpha\beta\gamma\delta} E^{\alpha\beta\gamma\delta},$$

alongside with the easily checkable identity,

$$(2 R^\sigma_\alpha E_{\sigma\beta\gamma\delta} + 3 R^{\sigma\tau}{}_{\alpha\beta} E_{\sigma\tau\gamma\delta}) E^{\alpha\beta\gamma\delta} \equiv 0.$$

1.3 The Noether Theorem: Space-Time Invariance

The famous Noether theorem describes a connection between symmetry properties of physical systems on the one hand, and conservation laws and concrete structure of conserved quantities on the other hand. Under the symmetry properties we understand invariance of the most universal characteristics of physical systems, the action integrals — or the corresponding Lagrangian densities —, under the symmetry transformations. The latter ones may represent merely space-time coordinates transformations (then the conserved quantities acquire the nature of either energy, linear and angular momenta, or their mixture). They may, however, also belong to generalized (gauge) transformations, being then connected with conservation of such quantities as the electric charge, or generalized charges; our considerations will be restricted in this book only to the coordinates transformations. One may consider separately those transformations which depend on finite number of parameters, or the continuous (infinite-parametric) ones. For the former ones, the *first Noether theorem* is formulated, while in the latter case, one speaks about the *second Noether theorem*. It is clear that the first theorem can be considered as a special case of the second one. In this text we discuss the case of general (infinite-parametric) space-time transformations only, thus the second Noether theorem is to be considered. The following exposition is a modified version of the Noether theorem given in (Mitskievich 1958, 1969), with the same general conclusions as in these earlier versions.

Notice first of all that invariance (scalar property) of the action integral S (written for an *arbitrary* space-time region) under general transformations of coordinates, is completely equivalent to the scalar density property of the corresponding Lagrangian density, $\mathfrak{L}$:

$$\mathfrak{L}'(x') = |J|^{-1}\mathfrak{L}(x), \tag{1.3.1}$$

where J is the Jacobian of the transformation under consideration[14]. When we write here $\mathfrak{L}'(x')$, we mean that all arguments on which the Lagrangian density depends, are functions of x'^{μ}; moreover, the form of these arguments (which are in general some tensor quantities and their first derivatives, as well as the Christoffel symbols) corresponds to the system of coordinates $\{x'\}$, and not to $\{x\}$ (the immediate prime at $\mathfrak{L}$).

As to the arguments, let them be the field potentials, A_a, a being a collective index consisting of a set of individual indices, such as the coordinated tensorial ones, or tetrad

---

[14] The absolute value in (1.3.1), $|J|$, has its origin in the fact that under general coordinates transformations the very sign of integral over a four-dimensional volume, $\int_\Omega$, is transformed as

$$\left[\int_\Omega\right]' = \int_{\Omega'} = \int_\Omega \operatorname{sign} J.$$

The sign function of Jacobian, $\operatorname{sign} J := J/|J|$, appears here since in this transformation a standard form of the integral (with upper limits greater than the lower ones) at each stage is used, while when the Jacobian J changes its sign (*i.e.*, $\operatorname{sign} J = -1$), a simple substitution of the transformed coordinates into the symbol of multiple integration yields an odd number of single integral symbols in it changing their limits to upper ones smaller than lower ones. Of course, the Jacobian can change its sign only in the whole region of integration: if it becomes equal to zero somewhere, the transformation will be degenerate, and the corresponding coordinates, unacceptable.

indices (in general, including spinor or Yang–Mills-type indices). The reader should remember that we consider here the metric description of geometry (and gravitational field) only, while in the spinor field theory one has to take for independent gravitational variables the tetrad field components, thus requiring some minor modifications of our approach, leading however to more cumbersome expressions and calculations we are trying to avoid in this book. Moreover, it will be helpful to introduce components of the Christoffel symbols as individual objects (which may call to mind the Palatini variational formalism in deduction of the gravitational field equations); such an approach will indeed simplify the structure of our expressions and equations. Thus we split here the field potentials into the gravitational metric part, $g_{\mu\nu}$, the gravitational connection part, $\Gamma^{\alpha}_{\beta\gamma}$, and that of all other fields, $A_{a'}$, — these latter fields are often treated as the only physical ones, but we shall always consider the gravitational field as one of physical fields too (endowed naturally with its own individuality). This splitting means

$$\{A_a\} = \{g_{\mu\nu},\ \Gamma^{\alpha}_{\beta\gamma},\ A_{a'}\},\ \text{or}\ \{a\} = \{\ _{\mu\nu}\ ^{\alpha}_{\beta\gamma}\ _{a'}\}. \qquad (1.3.2)$$

Concerning the transformations of coordinates, we shall consider only those which are reducible to infinitesimal ones (i.e., we shall speak about the connected component of the identity transformation). In this case it is always possible to recover finite transformations through integration of the infinitesimal ones, but of course not such as inversions. Thus we may write

$$x'^{\mu} = x^{\mu} + \epsilon \xi^{\mu}, \qquad \xi^{\mu} = \xi^{\mu}(x), \qquad (1.3.3)$$

ϵ being an infinitesimal constant, and ξ an arbitrary differentiable vector field. The standard transformation coefficients of tensor components by virtue of infinitesimality read

$$\frac{\partial x'^{\mu}}{\partial x^{\nu}} = \delta^{\mu}_{\nu} + \epsilon \xi^{\mu}{}_{,\nu}, \qquad \frac{\partial x^{\mu}}{\partial x'^{\nu}} = \delta^{\mu}_{\nu} - \epsilon \xi^{\mu}{}_{,\nu}, \qquad (1.3.4)$$

(up to the first order of magnitude terms).

Let us introduce a new operation, δ, (not to be confused with a variation in the action principle, or change of a tensor in a parallel transportation):

$$\delta A_a := A'_a(x') - A_a(x) =: \epsilon A_a|^{\tau}_{\sigma} \xi^{\sigma}{}_{,\tau} \qquad (1.3.5)$$

(also up to the first order terms). The ultimate right-hand part of this definition (which one has to read from the right to the left according to "=:" notation) is applicable only to tensors and tensor densities. If we consider the left-hand part (that with ":=") as applicable to absolutely all objects, it takes for the Christoffel symbols the form

$$\delta \Gamma^{\lambda}_{\mu\nu} = \Gamma'^{\lambda}_{\mu\nu}(x') - \Gamma^{\lambda}_{\mu\nu}(x) = \epsilon \left[\Gamma^{\lambda}_{\mu\nu}|^{\tau}_{\sigma} \xi^{\sigma}{}_{,\tau} - \xi^{\lambda}{}_{,\mu\nu} \right], \qquad (1.3.6)$$

the last term being typical for connection coefficients. Obviously, the $\Gamma^{\lambda}_{\mu\nu}|^{\tau}_{\sigma}$ construction is here the same as it would be taken for a third-rank tensor:

$$\Gamma^{\lambda}_{\mu\nu}|^{\tau}_{\sigma} = \Gamma^{\tau}_{\mu\nu} \delta^{\lambda}_{\sigma} - \Gamma^{\lambda}_{\sigma\nu} \delta^{\tau}_{\mu} - \Gamma^{\lambda}_{\mu\sigma} \delta^{\tau}_{\nu}. \qquad (1.3.7)$$

It is easy to find that

$$\delta(A_{a,\alpha}) = (\delta A_a)_{,\alpha} - \epsilon A_{a,\beta}\xi^\beta{}_{,\alpha}, \qquad (1.3.8)$$

which means that the operations of δ and of partial differentiation do not mutually commute.

What does commute with the partial differentiation, is an operation which differs "slightly" from δ, this is namely $A'_a(x) - A_a(x)$ [cf. (1.3.5)]. Here it is important to trace in more detail the meaning of primes in both definitions; see also the footnote 6. What means the prime in x'^μ, is probably clear, but the prime at A in $A'_a(x')$ or especially in $A'_a(x)$ presents a little more subtle subject. This prime means a change of the dependence of components of A_a on space-time coordinates. An elementary example is worth being considered here: Let the object A_a be simply a scalar function, $\Phi(x)$. The change to new coordinate system means $x^\mu = x^\mu(x')$, so that $\Phi(x) = \Phi(x(x'))$: scalars are such objects which do not change under any transformation of coordinates; however the very property of no change means then $\Phi(x) = \Phi'(x')$ since Φ represents now *another* function of the primed coordinates which is of course different from that of the old (non-primed) ones. Only such change of the form of the function guarantees preservation of the same numerical value of it at the same location which is labelled in both systems of coordinates by different collections of four numbers, $\{x^0, x^1, x^2, x^3\}$ and $\{x'^0, x'^1, x'^2, x'^3\}$. The new construction $A'_a(x) - A_a(x)$ coincides in fact (up to its sign) with the Lie derivative (1.2.21). The definition of a Lie derivative for tensors and tensor densities [and also for more general objects obeying homogeneous transformation laws like (1.3.5), without inhomogeneity terms of the type of the last one in (1.3.6)], is (1.2.20). Otherwise one can write

$$\pounds_\xi = \xi^\mu \partial_\mu - \epsilon^{-1}\delta. \qquad (1.3.9)$$

Obtention of (1.2.20) from (1.2.21) is proposed as a simple exercise for the reader, as well as a proof of automatic commutation of the operations $\pounds$ and ∂_μ,

$$\pounds_\xi(A_{a,\mu}) \equiv (\pounds_\xi A_a)_{,\mu} \qquad (1.3.10)$$

[cf. (1.3.8)].

Now, the Lagrangian density of the total system of fields is a function of all fields potentials and their first partial derivatives,

$$\mathfrak{L}_{\text{tot}} = \mathfrak{L}_{\text{tot}}(g_{\mu\nu}, \Gamma^\alpha_{\beta\gamma}, A_{a'}; g_{\mu\nu,\sigma}, \Gamma^\alpha_{\beta\gamma,\sigma}, A_{a',\sigma}). \qquad (1.3.11)$$

It is clear that the metric tensor can never be set equal to zero, as well as (in an arbitrary coordinates system) the components of the Christoffel symbols. However there is no such restrictions concerning the other fields, $A_{a'}$, which may exist or not exist depending of the physical nature of the system under consideration. Thus the function $\mathfrak{L}_{\text{tot}}$ makes sense also when A'_a vanish (together with their derivatives). We shall denote the resulting function as the purely gravitational Lagrangian density,

$$\mathfrak{L}_{\text{gr}} := \mathfrak{L}_{\text{tot}}(g_{\mu\nu}, \Gamma^\alpha_{\beta\gamma}, 0) \qquad (1.3.12)$$

(remember that this is by no means a dynamical relation: it simply represents our play with the arguments, not an interplay of physical fields, and it is a matter of a pure convention). This formal construction is however very convenient (as all good conventions are) for the end of classification and ordering of physical fields, beginning with the gravitational one. Then it makes sense also to introduce

$$\mathfrak{L}_\mathrm{f} := \mathfrak{L}_\mathrm{tot} - \mathfrak{L}_\mathrm{gr}, \qquad (1.3.13)$$

the Lagrangian density of all non-gravitational fields (all terms in this relation depend on all their legitimate arguments). However, $\mathfrak{L}_\mathrm{f}$ should not depend on the Christoffel symbols (this is an essential part of the Palatini formalism ideology intended for ensuring the metric nature of connection coefficients): if those are to be present in the structure of the non-gravitational fields Lagrangian density, they should be expressed explicitly through the metric tensor components and their partial derivatives, while the gravitational field Lagrangian density should not depend on the metric tensor derivatives which are to be incorporated exclusively in the Christoffel symbols (this pertains to the Palatini ideology too). Then for the field equations we have:
For the g field,

$$\frac{\delta \mathfrak{L}_\mathrm{tot}}{\delta g_{\mu\nu}} \equiv \frac{\delta \mathfrak{L}_\mathrm{gr}}{\delta g_{\mu\nu}} + \frac{\delta \mathfrak{L}_\mathrm{f}}{\delta g_{\mu\nu}} = 0; \qquad (1.3.14)$$

for the Γ field (this is simply the standard definition of the Christoffel symbols written in a dynamical fashion)

$$\frac{\delta \mathfrak{L}_\mathrm{f}}{\delta \Gamma^\lambda_{\beta\gamma}} \equiv 0, \quad \frac{\delta \mathfrak{L}_\mathrm{tot}}{\delta \Gamma^\lambda_{\beta\gamma}} \equiv \frac{\delta \mathfrak{L}_\mathrm{gr}}{\delta \Gamma^\lambda_{\beta\gamma}} = 0; \qquad (1.3.15)$$

and for the $A_{a'}$ field,

$$\frac{\delta \mathfrak{L}_\mathrm{g}}{\delta A_{a'}} \equiv 0, \quad \frac{\delta \mathfrak{L}_\mathrm{tot}}{\delta A_{a'}} \equiv \frac{\delta \mathfrak{L}_\mathrm{f}}{\delta A_{a'}} = 0. \qquad (1.3.16)$$

In (1.3.14) both Lagrangian densities ($\mathfrak{L}_\mathrm{gr}$ and $\mathfrak{L}_\mathrm{f}$) equally participate: they both depend on the metric coefficients. In (1.3.15) we have in fact $\mathfrak{L}_\mathrm{gr}$ only. In (1.3.16) $\mathfrak{L}_\mathrm{f}$ is present only by virtue of the definition of $\mathfrak{L}_\mathrm{gr}$.

Both Lagrangian densities, $\mathfrak{L}_\mathrm{gr}$ and $\mathfrak{L}_\mathrm{f}$, are obviously scalar densities, as the total Lagrangian density, $\mathfrak{L}_\mathrm{tot}$, was from the very beginning supposed to be. Therefore all the strong relations of Noether's theorem should be automatically written for each of them. Under *strong* relations we understand those which depend on the invariance properties only and not on the dynamical equations. Those which depend on both of them, are *weak* relations which involve also physical content, not only formal mathematical one, so that we consider weak relations as much more informative ones.[15] In order to come from the strong to weak relations, one has to make a substantial use of the field equations (1.3.14)–(1.3.16).

After these introductory remarks we may pass to the Noether theorem proper. The scalar density property of the Lagrangian density of a physical (field) system under arbitrary coordinates transformations, corresponds to the general covariance of the laws of dynamics

---
[15] *Cf.* the treatment of the properties of strength and weakness in the philosophy of Zen.

for these fields, which one obtains through the variational principle applied to the action integral for that Lagrangian density. Since variations of the field potentials should vanish at the space-time boundaries of the integration region, this scalar density property is of course somewhat excessive, but it is at least desirable for obtaining conserved quantities with good transformation properties. Thus we postulate the scalar density behavior of the Lagrangian density.[16] Thus we shall use the transformation law (1.3.1). In terms of the Lie derivative, $\pounds$, it may be obviously rewritten as

$$\pounds_\xi \mathfrak{L} - (\mathfrak{L}\xi^\alpha)_{,\alpha} = 0, \qquad (1.3.17)$$

where the partial divergence may be changed to the covariant one by virtue of the density property of $\mathfrak{L}$. We see that this equation (expressing the scalar density property of $\mathfrak{L}$ and nothing else) has explicitly covariant form.

In conformity with the above-discussed statements,

$$\mathfrak{L}(A_a, A_{a,\alpha}) \equiv \mathfrak{L}(g_{\mu\nu}, \Gamma^\lambda_{\beta\gamma}, A_{a'}; g_{\mu\nu,\alpha}, \Gamma^\lambda_{\beta\gamma,\alpha}, A_{a',\alpha}). \qquad (1.3.18)$$

When substituted into the relation (1.3.17) and due to the commutativity property (1.3.10), this expression yields

$$\frac{\delta \mathfrak{L}}{\delta g_{\mu\nu}} \pounds_\xi g_{\mu\nu} + \frac{\delta \mathfrak{L}}{\delta \Gamma^\lambda_{\beta\gamma}} \pounds_\xi \Gamma^\lambda_{\beta\gamma} + \frac{\delta \mathfrak{L}}{\delta A_{a'}} \pounds_\xi A_{a'} + \left(\frac{\partial \mathfrak{L}}{\partial g_{\mu\nu,\alpha}} \pounds_\xi g_{\mu\nu} \right.$$

$$\left. + \frac{\partial \mathfrak{L}}{\partial \Gamma^\lambda_{\beta\gamma,\alpha}} \pounds_\xi \Gamma^\lambda_{\beta\gamma} + \frac{\partial \mathfrak{L}}{\partial A_{a',\alpha}} \pounds_\xi A_{a'} - \mathfrak{L}\xi^\alpha \right)_{,\alpha} = 0. \qquad (1.3.19)$$

(Here we have performed the same rearrangements which are typical to the standard ones for the variational principle in field theory.) Now we have to take into account the facts that

$$\pounds_\xi g_{\mu\nu} = g_{\mu\nu,\sigma}\xi^\sigma + 2g_{\sigma(\mu}\xi^\sigma_{,\nu)}, \qquad (1.3.20)$$

$$\pounds_\xi \Gamma^\lambda_{\beta\gamma} = \Gamma^\lambda_{\beta\gamma,\sigma}\xi^\sigma - \Gamma^\lambda_{\beta\gamma}|^\tau_\sigma \xi^\sigma_{,\tau} + \xi^\lambda_{,\beta\gamma}, \qquad (1.3.21)$$

$$\pounds_\xi A_{a'} = A_{a',\sigma}\xi^\sigma - A_{a'}|^\tau_\sigma \xi^\sigma_{,\tau} \qquad (1.3.22)$$

as well as the relations (1.3.15) and (1.3.16) (however, sometimes we shall find it to be more convenient to return from separate summations of components of $g_{\mu\nu}$, $\Gamma^\lambda_{\sigma\tau}$ and $A_{a'}$ to components of the former object, A_a, adding where necessary a specific term due to $\Gamma^\lambda_{\sigma\tau}$, and without mentioning to which types of fields belongs $\mathfrak{L}$). Insertion of these expressions into (1.3.19) with subsequent rearrangement of all terms so that the only expression outside of the divergence should be proportional to ξ (and not to its derivatives), yields

$$\mathfrak{A}_\alpha \xi^\alpha = (\mathfrak{U}^\alpha_\sigma \xi^\sigma + \mathfrak{M}^{\alpha\tau}_\sigma \xi^\sigma_{,\tau} + \mathfrak{N}^{\alpha\tau\beta}_\sigma \xi^\sigma_{,\beta\tau})_{,\alpha}, \qquad (1.3.23)$$

---

[16] In fact, this does not hamper such procedures as truncation of divergence terms in the Lagrangians, e.g. when passing from Hilbert's to Einstein's representations of those, if the corresponding contribution to the boundary conditions is taken into account.

where new notations are used,

$$\mathfrak{A}_\alpha := \frac{\delta\mathfrak{L}}{\delta g_{\mu\nu}} g_{\mu\nu,\alpha} + \left(\frac{\delta\mathfrak{L}}{\delta g_{\mu\nu}} g_{\mu\nu}|_\alpha^\tau \right)_{,\tau} ; \qquad (1.3.24)$$

$$\mathfrak{U}_\sigma^\alpha := \frac{\delta\mathfrak{L}}{\delta g_{\mu\nu}} g_{\mu\nu}|_\sigma^\alpha - \mathfrak{t}_\sigma^\alpha, \qquad (1.3.25)$$

where

$$\mathfrak{t}_\sigma^\alpha := \frac{\partial\mathfrak{L}}{\partial A_{a,\alpha}} A_{a,\sigma} - \mathfrak{L}\delta_\sigma^\alpha; \qquad (1.3.26)$$

$$\mathfrak{M}_\sigma^{\alpha\tau} := \frac{\partial\mathfrak{L}}{\partial A_{a,\alpha}} A_a|_\sigma^\tau; \qquad (1.3.27)$$

$$\mathfrak{N}_\sigma^{\alpha\tau\beta} := -\frac{\partial\mathfrak{L}}{\partial \Gamma^\sigma_{\tau\beta,\alpha}} \qquad (1.3.28)$$

(the Noether densities). The last of these quantities is obviously symmetric (by the definition) in its two last upper indices, $\mathfrak{N}_\sigma^{\alpha\tau\beta} \equiv \mathfrak{N}_\sigma^{\alpha\beta\tau}$.

It is clear that the equation (1.3.23) expresses the condition on the Lagrangian, $\mathfrak{L}$, to be a scalar density under arbitrary transformations of coordinates (in the language of infinitesimal transformations, for arbitrary vector field ξ). However this condition can be formulated also irrespective of the choice of the field ξ, thus being connected directly with the intrinsic structure of the Lagrangian. This may be done after performing differentiations in (1.3.23),

$$(\mathfrak{U}_{\alpha,\beta}^\beta - \mathfrak{A}_\alpha)\xi^\alpha + (\mathfrak{U}_\sigma^\alpha + \mathfrak{M}_{\sigma,\beta}^{\beta\alpha})\xi^\sigma_{,\alpha} + (\mathfrak{M}_\sigma^{\alpha\tau} + \mathfrak{N}_{\sigma,\beta}^{\beta\tau\alpha})\xi^\sigma_{,\alpha,\tau} + \mathfrak{N}_\sigma^{\alpha\tau\beta}\xi^\sigma_{,\alpha,\tau,\beta} = 0. \qquad (1.3.29)$$

By virtue of the arbitrariness of all four components of ξ^μ, as well as of all their derivatives (with the obvious exception of the symmetry properties due to repeated differentiations), this equation splits into a system of multi-component equations which we shall call the *Noether relations*:

$$\mathfrak{A}_\alpha = \mathfrak{U}_{\alpha,\beta}^\beta, \qquad (1.3.30)$$

$$\mathfrak{U}_\sigma^\alpha + \mathfrak{M}_{\sigma,\beta}^{\beta\alpha} = 0, \qquad (1.3.31)$$

$$\mathfrak{M}_\sigma^{(\alpha\tau)} + \mathfrak{N}_{\sigma,\beta}^{\beta\tau\alpha} = 0, \qquad (1.3.32)$$

$$\mathfrak{N}_\sigma^{(\alpha\tau\beta)} = 0. \qquad (1.3.33)$$

Notice however that there exists an absolute identity

$$\frac{\partial\mathfrak{L}}{\partial x^\alpha} \equiv \frac{\delta\mathfrak{L}}{\delta A_a} A_{a,\alpha} + \left(\frac{\partial\mathfrak{L}}{\partial A_{a,\beta}} A_{a,\alpha} \right)_{,\beta}, \qquad (1.3.34)$$

independent of any invariance properties of $\mathfrak{L}$, which leads to a similar absolute identity property of (1.3.30).

Now, the Noether relations (1.3.31) – (1.3.33) yield

$$\mathfrak{U}^\alpha_{\sigma,\alpha} = 0, \tag{1.3.35}$$

being obviously equivalent to vanishing of $\mathfrak{A}_\alpha$:

$$\mathfrak{A}_\alpha = 0. \tag{1.3.36}$$

This simple property is of immense importance: it leads directly to definition of the symmetric energy-momentum tensor as well as to its covariant conservation law. In order to come to these conclusions, first we have to consider the structure of Lagrangian densities of physical fields and the corresponding field equations.

Let us consider from this point of view the relation (1.3.36). In fact, it reads[17]

$$\frac{\delta \mathfrak{L}}{\delta g_{\mu\nu}} g_{\mu\nu,\alpha} + \left(\frac{\delta \mathfrak{L}}{\delta g_{\mu\nu}} g_{\mu\nu}|^\beta_\alpha \right)_{,\beta} = 0. \tag{1.3.37}$$

Now we observe that the quantity $\frac{\delta \mathfrak{L}}{\delta g_{\mu\nu}} g_{\mu\nu}|^\beta_\alpha$ has the transformational properties of a rank-two tensor density (once co- and once contravariant, which corresponds to the indices α and β), while $\frac{\delta \mathfrak{L}}{\delta g_{\mu\nu}}$ is also a rank-two tensor density, but completely contravariant one (indices μ and ν). Thus we may replace the partial derivatives by covariant ones, inserting necessary terms with the Christoffel symbols for compensation:

$$\frac{\delta \mathfrak{L}}{\delta g_{\mu\nu}} \left(g_{\mu\nu;\alpha} + g_{\sigma\nu} \Gamma^\sigma_{\mu\alpha} + g_{\mu\sigma} \Gamma^\sigma_{\nu\alpha} \right)$$

$$+ \left(\frac{\delta \mathfrak{L}}{\delta g_{\mu\nu}} g_{\mu\nu}|^\beta_\alpha \right)_{;\beta} + \frac{\delta \mathfrak{L}}{\delta g_{\mu\nu}} g_{\mu\nu}|^\sigma_\tau \left(\Gamma^\tau_{\alpha\beta} \delta^\beta_\sigma - \Gamma^\beta_{\sigma\beta} \delta^\tau_\alpha - \Gamma^\beta_{\beta\alpha} \delta^\tau_\sigma \right) = 0. \tag{1.3.38}$$

We made here use of the transformation law for a tensor density (of weight one),

$$\left(\frac{\delta \mathfrak{L}}{\delta g_{\mu\nu}} g_{\mu\nu}|^\beta_\alpha \right)\bigg|^\sigma_\tau = \frac{\delta \mathfrak{L}}{\delta g_{\mu\nu}} g_{\mu\nu}|^\sigma_\alpha \delta^\beta_\tau - \frac{\delta \mathfrak{L}}{\delta g_{\mu\nu}} g_{\mu\nu}|^\beta_\tau \delta^\sigma_\alpha - \frac{\delta \mathfrak{L}}{\delta g_{\mu\nu}} g_{\mu\nu}|^\beta_\alpha \delta^\sigma_\tau, \tag{1.3.39}$$

the last term corresponding to the density property. In (1.3.38) one has to take into account that $g_{\mu\nu;\alpha} \equiv 0$, as well as that $g_{\mu\nu}|^\tau_\sigma = -2g_{\sigma(\mu} \delta^\tau_{\nu)}$. Then one gets simply

$$\mathfrak{T}^\beta_{\alpha;\beta} = 0, \tag{1.3.40}$$

where

$$\mathfrak{T}^\beta_\alpha := \frac{\delta \mathfrak{L}}{\delta g_{\mu\nu}} g_{\mu\nu}|^\beta_\alpha \equiv \frac{\delta \mathfrak{L}}{\delta g^{\mu\nu}} g^{\mu\nu}|^\beta_\alpha. \tag{1.3.41}$$

---

[17]The terms which included Γ, all vanished for any of the three Lagrangian densities (for $\mathfrak{L}_{\text{tot}}$ vanished, of course, all terms, thus limiting us to the other two cases; a similar situation took place with respect to $A_{a'}$). So the terms with $g_{\mu\nu}$ are the only worth being considered, and we omitted the subscripts (gr or f).

But usually a rank-two tensor, and not its density, is considered,

$$T^\beta_\alpha = (-g)^{-1/2} \mathfrak{T}^\beta_\alpha, \tag{1.3.42}$$

or

$$T^{\mu\nu} = -\frac{2}{\sqrt{-g}} \frac{\delta \mathfrak{L}}{\delta g_{\mu\nu}}, \quad T_{\mu\nu} = +\frac{2}{\sqrt{-g}} \frac{\delta \mathfrak{L}}{\delta g^{\mu\nu}}. \tag{1.3.43}$$

Then the relation (1.3.40) takes form

$$T^{\mu\nu}{}_{;\nu} = 0. \tag{1.3.44}$$

This relation is generally called a *covariant* conservation law,[18] the conserved quantity $T^{\mu\nu}$ being the stress-energy tensor (or: energy-momentum tensor), if in its definition $\mathfrak{L}_f$ is used. If this tensor is built with the help of $\mathfrak{L}_{gr}$, it represents (up to a constant factor) the conservative Einstein tensor (left-hand side of Einstein's equations: $R^{\mu\nu} - \frac{1}{2} g^{\mu\nu} R$), and the corresponding conservation law (vanishing of the covariant divergence) represents the repeatedly (twice) contracted Bianchi identities. Sometimes, both laws are treated as "Bianchi identities". Finally, $T^{\mu\nu}_{tot} \equiv T^{\mu\nu}_f + T^{\mu\nu}_{gr} = 0$ is nothing else than Einstein's equations if as $\mathfrak{L}_{gr}$, the standard Hilbertian Lagrangian of gravitational field is used.

As an example, consider now a special case of the stress-energy tensor — that of the electromagnetic field. The corresponding Lagrangian takes the form

$$\mathfrak{L}_{em} = -\frac{\sqrt{-g}}{16\pi} F^{\mu\nu} F_{\mu\nu}. \tag{1.3.45}$$

First, let us distribute the metric tensor determinant equally between both of F:

$$\mathfrak{L}_{em} = -\frac{1}{16\pi} F_{\kappa\mu} F_{\lambda\nu} \mathcal{G}^{\kappa\lambda} \mathcal{G}^{\mu\nu}, \tag{1.3.46}$$

where $\mathcal{G}^{\mu\nu} := (-g)^{1/4} g^{\mu\nu}$. Then, by an obvious extension of the definition (1.3.43),

$$\mathfrak{T}^\beta_\alpha := \frac{\partial \mathfrak{L}}{\partial \mathcal{G}^{\mu\nu}} \mathcal{G}^{\mu\nu}|^\beta_\alpha. \tag{1.3.47}$$

$\mathcal{G}^{\mu\nu}$ is here a contravariant tensor density of weight $\frac{1}{2}$, thus $\mathcal{G}^{\mu\nu}|^\beta_\alpha = \mathcal{G}^{\beta\nu} \delta^\mu_\alpha + \mathcal{G}^{\mu\beta} \delta^\nu_\alpha - \frac{1}{2} \mathcal{G}^{\mu\nu} \delta^\beta_\alpha$. Hence,

$$T^\beta_\alpha = -\frac{1}{8\pi\sqrt{-g}} F_{\kappa\mu} F_{\lambda\nu} \mathcal{G}^{\kappa\lambda} \left(\mathcal{G}^{\beta\nu} \delta^\mu_\alpha + \mathcal{G}^{\mu\beta} \delta^\nu_\alpha - \frac{1}{2} \mathcal{G}^{\mu\nu} \delta^\beta_\alpha \right), \tag{1.3.48}$$

i.e.

$$T^\beta_\alpha = -\frac{1}{16\pi} \left(4 F_{\alpha\nu} F^{\beta\nu} - F_{\mu\nu} F^{\mu\nu} \delta^\beta_\alpha \right). \tag{1.3.49}$$

---

[18] Since the covariant divergence here cannot be globally reduced to the partial one, the Gauss theorem cannot be applied to the integral of (1.3.44) over a four-dimensional (in a vulgar but rigorous enough sense, also three-dimensional) volume. See important comments just before the relation (1.3.54).

It is worth mentioning that the well known fact of identical vanishing of trace of the electromagnetic stress-energy tensor is now obvious not only from this last (standard) expression, but already from the expression of $\mathcal{G}^{\mu\nu}|_\alpha^\beta = \mathcal{G}^{\beta\nu}\delta_\alpha^\mu + \mathcal{G}^{\mu\beta}\delta_\alpha^\nu - \frac{1}{2}\mathcal{G}^{\mu\nu}\delta_\alpha^\beta$: its trace (with respect to the pair of indices α and β) readily vanishes.

Let us reconsider now the relation (1.3.23) taking into account (1.3.36); it reads then

$$(\mathfrak{U}_\sigma^\alpha \xi^\sigma + \mathfrak{M}_\sigma^{\alpha\tau}\xi^\sigma{}_{,\tau} + \mathfrak{N}_\sigma^{\alpha\tau\beta}\xi^\sigma{}_{,\beta,\tau})_{,\alpha} = 0. \tag{1.3.50}$$

Note that this relation has a scalar density nature; this fact can be traced from the early step, (1.3.17), and then (1.3.19) (remember also that the quantity $\mathfrak{A}_\alpha$ represents a covector density). Thus, under general transformations of coordinates, the expression inside the parentheses,

$$\mathfrak{w}^\alpha := \mathfrak{U}_\sigma^\alpha \xi^\sigma + \mathfrak{M}_\sigma^{\alpha\tau}\xi^\sigma{}_{,\tau} + \mathfrak{N}_\sigma^{\alpha\tau\beta}\xi^\sigma{}_{,\beta,\tau}, \tag{1.3.51}$$

behaves as a vector density. Then the relation (1.3.50) can be interpreted as a general covariant differential conservation law,[19]

$$\mathfrak{w}^\alpha{}_{;\alpha} \equiv \mathfrak{w}^\alpha{}_{,\alpha} = 0. \tag{1.3.52}$$

Here the identical equality of covariant divergence to the partial one, is obvious from the vector density property of $\mathfrak{w}^\alpha$: one simply has to use the general expression of a covariant derivative,

$$T_{a;\alpha} = T_{a,\alpha} + T_a|_\tau^\sigma \Gamma_{\sigma\alpha}^\tau. \tag{1.3.53}$$

The reduction of covariant divergence to a partial one in (1.3.52), is of great importance: it allows to apply the Gauss theorem when this law is integrated over a four-dimensional region. Then a usual scalar integral conservation law emerges:

$$W_+ - W_- = -\int_{V_{\text{lat}}} \mathfrak{w}^\alpha dS_\alpha, \tag{1.3.54}$$

where

$$W_\pm := \int_{V_\pm} \mathfrak{w}^\alpha dS_\alpha, \tag{1.3.55}$$

dS_α being space-like hypersurface element of the final/initial moment of time ($\pm$) (we consider the integration region as interior of a hypercylinder in space-time bounded below and above by constant time hyperplanes), while V_{lat} represents (time-like) lateral cylindrical hypersurface of the boundary. The right-hand side term in (1.3.54) describes (time-integrated) flux of the conserved quantity through the spatial boundary, and the left-hand side terms, the change (during this time interval) of the W-contents of the three-dimensional volume surrounded by this lateral surface. The marvel of this result is that it is general covariant at all its stages, and it is related to all possible conserved quantities (energy, momentum, angular momentum), if the physical system under consideration possesses necessary invariance properties.

---

[19]General covariance reduces to the tensorial one only as to its special case, thus the former one is really more general. In fact, the conservation law (1.3.52) possesses the scalar density property.

A proper interpretation of this fact is however not as elementary as one could expect at the first glance. The simplest case corresponds to the flat Minkowski space-time in which there are ten (the maximal number of) Killing vectors which correspond in this case to: four Cartesian directions (or pure translations); three spatial (trigonometric) rotations; and three boosts (hyperbolic "rotations"). Then one has to connect with the translational isometries the four transformation vector fields $\xi^\mu = \delta^\mu_\sigma$, where σ takes fixed values 0, 1, 2, and 3. The temporal translation ($\sigma = 0$) is related to energy conservation, the other three values, to conservation of the corresponding components of linear momentum (impulse). Rotations and boosts are described by $\epsilon \xi^\mu \Rightarrow \omega^\mu{}_\nu x^\nu$, where a set of infinitesimal constants, $\omega^\mu{}_\nu$, is introduced instead of only one former factor ϵ. Since we consider at this stage the Minkowski space-time case, Cartesian coordinates are singled out as preferred ones, and we may choose the transformation between x^μ and x'^μ as connecting two such systems of coordinates (only in this case one can speak about the "true" rotations and boosts). These rotations and boosts transformations are then (up to the first order of smallness) orthogonal transformations only if $\omega_{\mu\nu} = -\omega_{\nu\mu}$, but otherwise ω's are arbitrary. Their exclusion yields to the standard laws of conservation, the corresponding conserved quantities being components of angular momentum (for spatial rotations) and those of the centre-of-mass characteristics (for boosts).

In the general case when there are no Killing vectors, one cannot choose any preferred form for the vectors which could be interpreted (at least in a restricted region) as generators of translations and rotations. However the conservation law (1.3.52) still holds exactly for an arbitrary (continuous) choice of ξ in the vector density $\mathfrak{w}^\alpha$ (1.3.51), so that we have in fact a continuum of conserved quantities and conservation laws, now without a possibility to decompose them with respect to some basis, as it was the case in the (flat) Minkowski space-time (a basis, in the sense of Killing vectors, of course). Contrary to a relatively widespread point of view of absence of any well defined conservation laws in such maximally inhomogeneous space-times, the whole variety of conservations survives even there, but now there exists no criterion to classify the corresponding transformations (special choices of the arbitrary vector field ξ) — it is impossible to single out of this continuum of quantities, specifically those which are the most customary for us, that is, energy and linear and angular momenta. This situation is neither bad nor good one, it is simply a real and normal situation. In such a maximally inhomogeneous space-time, a transformation which we interpret (e.g.) as a translation at one point, becomes a mixture of translations and rotations (including hyperbolic ones) at the neigbouring points, probably even a pure rotation or boost at some more distant point, this situation being absolutely incurable. But this represents a *normal* state of affairs for the general case of Riemannian geometry. Only when there exists (at least) one Killing vector, it becomes possible to baptize the corresponding infinitesimal transformation, conserved quantity, and the conservation law, giving them some name(s). This name, however, will represent — even in this case of existence of an isometry — a mere convention. This latter fact can be somewhat clarified by an example. Say, we consider a space-time of some constant curvature, different from zero. In such a space-time, like it was in the Minkowski case, there are all the ten possible Killing vectors

— the maximal number for the dimension of four. But, in contrast with the Minkowski space-time, no one of these Killing vectors does now describe a *pure* transformation, since it changes along its own integral curve from (say) translation to rotation, and so on. Only in exceptional (though maybe important for us) cases, we can encounter a pure translation (in time, or in a spatial direction), or a pure rotation, or a pure boost. In the Kerr space-time as another example, we have two Killing vectors, one of them (the time-like one) describing a temporal translation, and another (the spatial, ∂_ϕ) — a rotation about the z axis. In this special case, we may single out of the whole variety of the conserved quantities, only two of them,— energy and one (z-) component of angular momentum, together with the corresponding conservation laws. All other conserved quantities and conservation laws do certainly exist, but they continue to exist as one whole bulk.

Another remark which is worth being mentioned, is related to the fact that the customary interpretation of conserved quantities as (e.g.) vector ones, is admissible in the flat (Minkowski) world *and* for Cartesian coordinates only. This means that (as it was the case for interpretation of the four coordinates, x^μ) each "component" of such a "vector" is in fact a *scalar function*. This fact is rather obvious though we all are conditioned not to acknowledge it.

It is clear that $\mathfrak{t}_\nu^\mu$ does not behave as a tensor density (and certainly not as a tensor) under arbitrary transformations of coordinates; the term "pseudotensor" maybe somewhat confusing since it calls to mind pseudo-objects (such as pseudoscalars which change sign under inversions of odd numbers of coordinates) with which it has nothing in common. The tradition has however a strong vitality in this case, so we don't feel it possible to change the terminology. The difference between $\mathfrak{t}_\sigma^\alpha$ and the density of T_σ^α (i.e. $\mathfrak{T}_\sigma^\alpha$) is described due to (1.3.25), (1.3.26) and (1.3.41) as $\mathfrak{U}_\sigma^\alpha$,

$$\mathfrak{T}_\sigma^\alpha = \mathfrak{t}_\sigma^\alpha + \mathfrak{U}_\sigma^\alpha. \tag{1.3.56}$$

Here the left-hand side represents a tensor density, while in the right-hand side there are two non-tensor quantities (however their exact transformation laws under general transformations of coordinates are easily obtainable). These latter quantities certainly behave as tensor densities under linear transformations (such objects are called *affine tensor densities*); for that reason the just mentioned name of pseudotensor is in fact used. The tensor density proper, $\mathfrak{T}_\sigma^\alpha$, is symmetric (one commonly applies the concept of symmetry when a quantity does not change when the indices are simply transposed, but it is applicable also in the case of a transposition which conserves the level — co- or contravariant — of these indices, thus $\mathfrak{T}^\alpha{}_\sigma = \mathfrak{T}_\sigma{}^\alpha$). The energy-momentum pseudotensor, $\mathfrak{t}_\sigma^\alpha$, possesses, in its turn, no such symmetry (only in the case when the energy-momentum of a scalar field is considered, it is in fact symmetric, but then the term $\mathfrak{U}_\sigma^\alpha$ vanishes identically). The same no symmetry property is characteristic to the other pseudotensor, $\mathfrak{U}_\sigma^\alpha$, which is said to symmetrize the energy-momentum pseudotensor $\mathfrak{t}_\sigma^\alpha$ to the symmetric tensor density $\mathfrak{T}_\sigma^\alpha$. The symmetrization mechanism was found in 1939 by F.J. Belinfante and in 1940 by L. Rosenfeld, many years after the discovery of the Noether theorem in 1918. Since no symmetry property of the canonical energy-momentum pseudotensor $\mathfrak{t}_\sigma^\alpha$ is due to possession of a non-zero spin by the corresponding field (remember that in the case of a spinless scalar field no such problem arises), the pseudotensor $\mathfrak{U}^\alpha_\sigma$ can be named the *spin part of energy-momentum*. In its turn, the Noether relation (1.3.31) reduces $\mathfrak{U}^\alpha_\sigma$ to a divergence of $\mathfrak{M}^{\beta\alpha}_\sigma$, the quantity which enters angular momentum density in the special relativistic limit for pure rotations. This is exactly the form in which symmetrization of the canonical energy-momentum pseudotensor was achieved by Belinfante and Rosenfeld. In the special relativistic limit, $\mathfrak{M}^{\beta\alpha}_\sigma$ (after the corresponding antisymmetrization due to antisymmetry of the infinitesimal factor $\omega^\mu{}_\nu$, preferably after lowering the first index, μ, with help of the Cartesian Minkowski metric tensor) is interpreted as the spin density (or proper angular momentum pseudotensor). Without such an antisymmetrization, $\mathfrak{M}^{\beta\alpha}_\sigma$ is better to be named generalized spin density. Quite naturally, this spin pseudotensor yields, after the second quantization procedure, just the standard spin expression for the corresponding field whose Lagrangian density is employed in building the pseudotensor. The last of the Noether densities, $\mathfrak{N}^{\beta\alpha\tau}_\sigma$, is called *bispin* density, and it is a tensor density proper, but its physical significance still is unclear.

1.4 The Noether Densities' Transformation Laws

Transformation laws of quantities like $\mathfrak{t}^\alpha_\sigma$ (which we call Noether densities), in particular, of the conserved quantities, are of great importance to adequately understand dynamical characteristics of physical objects. When Einstein (1916) introduced the concept of gravitational field energy, Bauer (1918) promptly showed that — in the flat space-time — in spherical coordinates the integral of this energy is not simply different from zero, but diverges, although the same integral is equal to zero in Cartesian coordinates. But it is clear that in the flat (Minkowski) space-time there is no gravitational field at all (the space-time is perfectly homogeneous there); moreover, passing from a Cartesian system to a spherical one, one does not introduce any relative *motion* of the corresponding frames, thus the transformation is performed within the framework of a fixed frame of reference, independently of the concrete definition of the latter concept. Therefore one cannot speak in this case about an appearance of anything similar to kinetic energy or of energy of the field of inertial forces: nothing of the kind is possible here. Nevertheless, 'the gravitational energy' (both its density and integral) in the form proposed by Einstein, suffers essential changes under such a 'harmless' transformation.

Since we traditionally consider energy as one of the most important characteristics of physical systems, it is worth subjecting it from the very beginning to certain strict conditions. These should be formulated on the basis of physical ideas; however, the latter ones are frequently written in integral sense only, *for the universe as a whole*. This fact reflects the role of special relativity in our thinking, but the situation described by special relativity, corresponds to a homogeneous geometry of the universe, in particular, to existence in this case of global Cartesian coordinates, as well as of an exclusive role of linear coordinate transformations acting between such Cartesian systems. In special relativity (and only *under linear transformations* even here) the integral energy and momentum, taken together, form a four-vector (with respect to the Lorentz transformations). However in general relativity it is impossible to introduce transformations which could be in general consistently analogous to the Lorentz ones, since in a curved universe there is no concept of a straight line, hence of a uniform rectilinear motion, especially when one means not a local transformation, or that along a single line, but a transformation in, at least, a finite four-dimensional region. If one considers an insular model of universe, all the matter in it would be concentrated in a limited region of space, leaving empty the infinite surroundings of it, so that one could look at this distribution 'from outside'. The observer is then located in a practically flat space-time, and can describe the system (universe) under consideration on such a flat background, *i.e.* in terms of special relativity. Though the really existing universe is at our disposal in a single edition only, our theory should be formulated so that it has to give a sensible description of any system, as long as we specify some well defined initial and boundary conditions. Therefore, to give an introduction of the problem, we shall begin with the energy-momentum as it should be considered by an 'outside' observer. This approach was formulated by Møller (1961a, 1961b).

Møller's Conditions

I. The canonical pseudotensor $\mathfrak{t}_\sigma^\alpha$ at an arbitrary world point x^μ should be an affine tensor density of weight +1, depending algebraically on fields' potentials and their first and second derivatives at the same point.

II. The quantity $\mathfrak{t}_\sigma^\alpha$ should satisfy the *weak* affine conservation law

$$\mathfrak{t}_{\sigma,\alpha}^\alpha = 0. \tag{1.4.1}$$

III. The energy density $\mathfrak{t}_0^\alpha$ should transform as density of a contravariant four-vector under general purely spatial transformations:

$$\mathfrak{t}_0'^\alpha = |J|^{-1} \mathfrak{t}_0^\beta \frac{\partial x'^\alpha}{\partial x^\beta}, \quad x'^i = x'^i(x^k), \quad x'^0 = x^0. \tag{1.4.2}$$

IV. Under linear transformation the integral energy-momentum 'vector'

$$P_\sigma = \int_\Sigma \mathfrak{t}_\sigma^\alpha dS_\alpha \tag{1.4.3}$$

should transform as a free four-covector; moreover, it should not suffer changes under transformations which reduce to the identity transformation at large spatial distances, but are arbitrary elsewhere.

V. In the centre-of-mass system, the energy-momentum 'vector' should take the form

$$(P_\sigma) = (M, 0, 0, 0). \tag{1.4.4}$$

These conditions are practically the same as Chr. Møller formulated them in 1961: we have only changed strong conservation law to weak one, and included an additional supposition — V. — (in fact, admitted by Møller himself). These five conditions are fairly natural, and the first two of them are automatically satisfied by all dynamical quantities following from Noether's theorem. The condition III. obviously guarantees invariance of integral energy in any (in particular, not isolated) volume under purely spatial transformations. It is designed to exclude the appearance of the already mentioned Bauer paradox, and in this sense the III. condition is necessary and sufficient. Only the conditions IV. and V. admit an insular model of universe, and this model justifies them perfectly. We would like to comment more on the invariability of P_σ under arbitrary transformations (of both spatial and temporal coordinates) coinciding with the identity transformation at great distances. Møller, speaking on linear transformations, meant the Lorentz transformations, as well as purely spatial rotations, of special relativity. All of them are well determined faraway from the physical system where space-time becomes practically flat. Near the system and inside it, where space-time is curved, the concept of a straight line loses its meaning, so that we cannot say which transformations should be considered to be linear there. Therefore in this interior region all transformations of four coordinates should be accepted indiscriminately without affecting the value of the integral (1.4.3) which should depend only on the asymptotic behaviour of the coordinates' transformations.

Let us now show that the Lagrangian's invariance guarantees satisfaction of the III. condition. We have already mentioned the fact that true scalar densities participate in expressions (1.3.17) and (1.3.19); moreover, we have put (1.3.36) (vanishing of (1.3.24), the expression in the first square brackets in (1.3.23), leading to the covariant conservation law (1.3.40) into the explicitly tensor form (that of a covector density). Since we have a scalar product of this expression with a contravariant vector, a scalar density emerges (this can be traced using (1.3.17) and (1.3.19) too). Therefore the divergence in (1.3.23) also yields a scalar density, and the expression in the second square brackets — in fact, $\mathfrak{w}^\alpha$ — is a contravariant vector density, in other words, it transforms as

$$\mathfrak{w}'^\alpha(x') = |J|^{-1}\frac{\partial x'^\alpha}{\partial x^\beta}\mathfrak{w}^\beta(x) \tag{1.4.5}$$

under the most general admissible coordinate transformations in the theory of relativity, $x'^\alpha = x'^\alpha(x^\mu)$ (not necessarily infinitesimal transformations already). We know, however, that ξ^σ is a contravariant vector, thus

$$\xi'^\sigma(x') = \frac{\partial x'^\sigma}{\partial x^\mu}\xi^\mu(x), \tag{1.4.6}$$

$$(\xi^\sigma_{,\tau})' = \frac{\partial x'^\sigma}{\partial x^\mu}\frac{\partial x^\nu}{\partial x'^\tau}\xi^\mu_{,\nu} + \left(\frac{\partial}{\partial x'^\tau}\frac{\partial x'^\sigma}{\partial x^\mu}\right)\xi^\mu, \tag{1.4.7}$$

$$(\xi^\sigma_{,\tau,\beta})' = \frac{\partial x'^\sigma}{\partial x^\mu}\frac{\partial x^\nu}{\partial x'^\tau}\frac{\partial x^\rho}{\partial x'^\beta}\xi^\mu_{,\nu,\rho} + \left(\frac{\partial x'^\sigma}{\partial x^\mu}\frac{\partial^2 x^\nu}{\partial x'^\tau \partial x'^\beta} + \frac{\partial x^\nu}{\partial x'^\tau}\frac{\partial}{\partial x'^\beta}\frac{\partial x'^\sigma}{\partial x^\mu}\right.$$
$$\left.+ \frac{\partial x^\nu}{\partial x'^\beta}\frac{\partial}{\partial x'^\tau}\frac{\partial x'^\sigma}{\partial x^\mu}\right)\xi^\mu_{,\nu} + \left(\frac{\partial^2}{\partial x'^\beta \partial x'^\tau}\frac{\partial x'^\sigma}{\partial x^\mu}\right)\xi^\mu. \tag{1.4.8}$$

Now we have to substitute these relations, together with the expression (1.3.51), in the transformation law (1.4.5); in the resulting equality the terms with ξ^μ, $\xi^\mu_{,\nu}$ and $\xi^\mu_{,\nu,\rho}$ should be put together separately. Like in Noether's theorem, it is necessary to treat here the vector ξ^μ and its derivatives locally as arbitrary functions of coordinates (not forgetting, of course, that the second derivative is symmetric in its lower indices). Then we easily arrive to the transformation laws

$$\mathfrak{U}^\nu_\mu = |J| \frac{\partial x^\nu}{\partial x'^\alpha} \left(\frac{\partial x'^\sigma}{\partial x^\mu} \mathfrak{U}'^\alpha_\sigma + \frac{\partial}{\partial x'^\tau} \frac{\partial x'^\sigma}{\partial x^\mu} \mathfrak{M}'^{\alpha\tau}_\sigma + \frac{\partial^2}{\partial x'^\beta \partial x'^\tau} \frac{\partial x'^\sigma}{\partial x^\mu} \mathfrak{N}'^{\alpha\tau\beta}_\sigma \right), \tag{1.4.9}$$

$$\mathfrak{M}^{\nu\lambda}_\mu = |J| \frac{\partial x^\nu}{\partial x'^\alpha} \left[\frac{\partial x'^\sigma}{\partial x^\mu} \frac{\partial x^\lambda}{\partial x'^\tau} \mathfrak{M}'^{\alpha\tau}_\sigma \right.$$
$$\left. + \left(\frac{\partial x^\lambda}{\partial x'^\tau} \frac{\partial}{\partial x'^\beta} \frac{\partial x'^\sigma}{\partial x^\mu} + \frac{\partial x^\lambda}{\partial x'^\beta} \frac{\partial}{\partial x'^\tau} \frac{\partial x'^\sigma}{\partial x^\mu} + \frac{\partial x'^\sigma}{\partial x^\mu} \frac{\partial^2 x^\lambda}{\partial x'^\beta \partial x'^\tau} \right) \mathfrak{N}'^{\alpha\tau\beta}_\sigma \right], \tag{1.4.10}$$

$$\mathfrak{N}^{\nu\lambda\rho}_\mu = |J| \frac{\partial x^\nu}{\partial x'^\alpha} \frac{\partial x^\lambda}{\partial x'^\tau} \frac{\partial x^\rho}{\partial x'^\beta} \frac{\partial x'^\sigma}{\partial x^\mu} \mathfrak{N}'^{\alpha\tau\beta}_\sigma. \tag{1.4.11}$$

These transformation laws are written as 'inverse' ones — expressing quantities in the unprimed coordinate system in terms of those in the primed system; in order to obtain the 'direct' transformations, one has to exchange primes and their absence, as well as take $|J|^{-1}$ instead of $|J|$. Since in the relation (1.3.56) $\mathfrak{T}^\alpha_\sigma$ is an exact tensor density, (1.4.9) readily yields the transformation law of the canonical energy-momentum pseudotensor:

$$\mathfrak{t}'^\nu_\mu = |J|^{-1} \frac{\partial x'^\nu}{\partial x^\alpha} \left(\frac{\partial x^\sigma}{\partial x'^\mu} \mathfrak{t}^\alpha_\sigma - \frac{\partial}{\partial x^\tau} \frac{\partial x^\sigma}{\partial x'^\mu} \mathfrak{M}^{\alpha\tau}_\sigma - \frac{\partial^2}{\partial x^\beta \partial x^\tau} \frac{\partial x^\sigma}{\partial x'^\mu} \mathfrak{N}^{\alpha\tau\beta}_\sigma \right). \tag{1.4.12}$$

We see that the bispin density simply is an exact tensor density of weight +1 and rank four; the generalized spin density together with the bispin density forms a geometric object; the canonical energy-momentum pseudotensor, generalized spin, and bispin together form a geometric object too.[20] Thus the Noether densities represent three geometric objects 'enveloping' one another in a kind of nesting dolls (the Russian матрёшки) manner; the simplest of these objects is the tensor density $\mathfrak{N}$.

It is easy to see that due to such simple transformation properties of $\mathfrak{U}$, $\mathfrak{M}$, and $\mathfrak{N}$, it is possible to combine them into purely tensorial quantities, using the Christoffel symbols and their first derivatives:

$$U^\alpha_\sigma = \frac{1}{\sqrt{-g}} \left[\mathfrak{U}^\alpha_\sigma + \mathfrak{M}^{\alpha\tau}_\omega \Gamma^\omega_{\sigma\tau} - \mathfrak{N}^{\alpha\tau\beta}_\omega \left(\frac{\partial \Gamma^\omega_{\sigma\beta}}{\partial x^\tau} + \Gamma^\epsilon_{\sigma\tau} \Gamma^\omega_{\epsilon\beta} - \Gamma^\epsilon_{\beta\tau} \Gamma^\omega_{\epsilon\sigma} \right) \right], \tag{1.4.13}$$

$$M^{\alpha\tau}_\sigma = \frac{1}{\sqrt{-g}} \left[\mathfrak{M}^{\alpha\tau}_\sigma + \mathfrak{N}^{\alpha\beta\gamma}_\omega \left(\Gamma^\tau_{\beta\gamma} \delta^\omega_\sigma - \Gamma^\omega_{\sigma\beta} \delta^\tau_\gamma - \Gamma^\omega_{\sigma\gamma} \delta^\tau_\beta \right) \right], \tag{1.4.14}$$

$$N^{\alpha\tau\beta}_\sigma = \frac{1}{\sqrt{-g}} \mathfrak{N}^{\alpha\tau\beta}_\sigma. \tag{1.4.15}$$

---

[20] A *geometric object* is a multy-component quantity whose components in a system $\{x'\}$ are linear combinations of those in $\{x\}$, the coefficients being (products of) partial derivatives of x' with respect to x and *vice versa* of different orders, with no other quantities participating in this transformation law.

They satisfy tensorial (Noether) identities with covariant differentiation and the curvature tensor in coefficients. The same truly tensor quantities U_σ^α, $M_\sigma^{\alpha\tau}$ and $N_\sigma^{\alpha\tau\beta}$ follow directly from Noether's theorem, if, beginning with the relations (1.3.23), one uses covariant derivatives of ξ^μ and not partial ones (but symmetrizing indices in repeated differentiations). This approach, however, does not give exact (affine) conservation laws involving the new quantities (1.4.13)–(1.4.15); see Mitskievich (1964b, 1965b).

Turning to the behavior of $\mathfrak{t}_0^\alpha$ under purely spatial coordinates transformations (1.4.2) in connection with the III. Møller's condition, we find from (1.4.12)

$$\mathfrak{t}_0^{\prime\nu} = |J|^{-1}\frac{\partial x^{\prime\nu}}{\partial x^\alpha}\mathfrak{t}_0^\alpha, \tag{1.4.16}$$

i.e. the exact transformation law of a vector density, as it was to be expected, since $\partial x^\sigma/\partial x^{\prime 0} = \delta_0^\sigma$ for these transformations of coordinates. We see that the III. Møller's condition is satisfied automatically due to the Lagrangian's invariance. Thus the reason of the Bauer paradox consisted in the non-invariant Lagrangian[21] used by Einstein in his energy-momentum pseudotensor.

It is interesting that the quantity $\mathfrak{M}_0^{\alpha 0}$ behaves under purely spatial transformations (1.4.2) as a vector density too. However it is worth to mention that, even in co-moving coordinates (with respect to the reference frame), the 'purely spatial coordinates transformations' do not exhaust the whole family of transformations which do not get out of the initial *reference frame*. The most general transformations of this kind (with coordinates co-moving with the reference frame) will be considered here later; they form the basis of the *chronometric invariants formalism* of Zel'manov, and when we finally abandon the fairly old-fashioned idea of merging coordinates with reference frames, this directly leads to the perfectly elegant monad approach presented, for example, in this book.

Mais revenons a nos moutons. When studying transformation behavior of the 'vector' P_σ from the point of view of an observer faraway from the physical system, *i.e.* in the language of special relativity, one has to distinguish between two possibilities in the sense of transformations themselves. The first possibility reduces to the mere transformations of coordinates, while the hypersurface over which the integration in (1.4.3) is performed, remains unaltered. If a faraway observer has chosen a hyperplane as such a hypersurface (according to the traditions of special relativity it is supposed to correspond to a certain moment of time in this initial frame), then this observer should take *another* hyperplane when passing to a new system of coordinates (now co-moving with this new reference frame) in a uniform and rectilinear motion with respect to the former frame, since the notion of simultaneity of this observer should, of course, change. Thus we naturally come to the conclusion that a transformation of coordinates not corresponding to one fixed reference frame, but without an appropriate change (hyperbolic rotation) of the hyperplane of integration in (1.4.3), could have only formal, but not a physical meaning. One has to perform simultaneously both transformation to coordinates co-moving with a new reference frame, *and* a change to another hyperplane of integration according to the new simultaneity relations in this new

---

[21]Einstein's Lagrangian differs from Hilbert's scalar density one by a non-scalar density divergence term.

frame. This is very specific for the special relativity theory; in a problem of general relativity, if we do not restrict our consideration to an insular model of Universe, the change of the hypersurface of integration will be even more sophisticated one, and in general one hardly has any hope to introduce in an exact form these hypersurfaces of simultaneity, especially if our space-time would admit gravitational lenses, to say nothing of wormholes. The alternative was noticed by Schmutzer (1964), who gave preference to the first variant (perhaps for some technical reasons). It is interesting to note that Møller himself did formulate his condition IV. precisely in the sense of the *second* possibility (the combined transformation). He did use in fact the typically three-dimensional expression for the integral quantity P_σ, even not relating dV to the other components of dS_α. From the condition I. (the property of affine scalar density) and taking into account the transformation of dS_α, the transformation law of the integral energy-momentum 'vector' under linear transformations of coordinates (in particular, under a Lorentz transformation), takes the form

$$P'_\sigma|_{\Sigma'} = \int_{\Sigma'} t'^\alpha_\sigma dS'_\alpha = \frac{\partial x^\tau}{\partial x'^\sigma} \int_{\Sigma'} t^\beta_\tau dS_\beta, \qquad (1.4.17)$$

where Σ' corresponds to another hypersurface than Σ in (1.4.3) — not only in the sense of coordinates in which the hypersurfaces are described. In these expressions only a coordinate transformation was performed, but the integration hypersurface remained unchanged. Let us compare this result with the integral over a new hypersurface (a new choice of 'simultaneity'), but for convenience rewriting the former coefficient of transformation:

$$\frac{\partial x^\tau}{\partial x'^\sigma} P_\tau|_\Sigma \equiv \frac{\partial x^\tau}{\partial x'^\sigma} \int_\Sigma t^\beta_\tau dS_\beta. \qquad (1.4.18)$$

The comparison yields

$$P'_\sigma|_{\Sigma'} - \frac{\partial x^\tau}{\partial x'^\sigma} P_\tau|_\Sigma = \frac{\partial x^\tau}{\partial x'^\sigma} \oint_\Sigma t^\beta_\tau dS_\beta = \frac{\partial x^\tau}{\partial x'^\sigma} \int_\Omega t^\beta_{\tau,\beta}(dx). \qquad (1.4.19)$$

Since the integration hypersurface goes to the space-like infinity where fields are absent by admission of the insular model for which the condition IV. is formulated, we did complement this integration hypersurface by some lateral sides, redefining the direction of normals to the hypersurface, and applying the Gauss theorem. This is possible only if integral over the (new) lateral hypersurface (having, of course, a space-like normal vector) rapidly enough tends to zero when this hypersurface goes to the spatial infinity. Møller's condition IV. consists exactly of this. Now, in view of the condition II. (the differential conservation law following from the Noether theorem), we have to conclude that

$$P'_\sigma|_{\Sigma'} = \frac{\partial x^\tau}{\partial x'^\sigma} P_\tau|_\Sigma. \qquad (1.4.20)$$

The condition IV. is thus satisfied.

To summarize: If the conditions I. and II. determine the necessary differential properties of conserved quantities of the energy-momentum type, the condition III. guarantees invariability of energy under purely spatial transformations (now, in the integral form), while the condition IV. connects transformation laws of the integral energy under Lorentz transformations, with the necessity of a change of simultaneity under these transformations (since now an integral, and not local, quantity is being considered). Finally, the condition V. is tantamount to postulation of the Galilei–Eötvös–Einstein equivalence principle.[22,23]

---

[22]The Galilei–Eötvös–Einstein equivalence principle states the equivalence of inertial and gravitational masses. This principle essentially holds in the non-relativistic approximation; this fact was noticed only quite recently, and the conclusions on its relativistic generalization were made: see Mitskievich (2003).

[23]However, regarding the last condition, there still is a fly (but who knows, maybe a butterfly) in the ointment: in $\mathfrak{w}^\alpha$, the vector field ξ could be identified with a monad τ only up to a scalar function factor.

Chapter 2

Reference Frames' Calculus

2.1 The Monad Formalism and Its Place in the Description of Reference Frames in Relativistic Physics

It would be completely erroneous to automatically identify the concepts of a system of coordinates and of a reference frame. Such an identification is often characteristic for those physicists who are accustomed to work in Cartesian coordinates and who, moreover, try to reduce (for the sake of "simplicity") problems — mainly, educational ones — to two-dimensional diagrams dealing with one temporal and another spatial, coordinates. But when more than one spatial dimension (plus extra one, temporal dimension) are considered, a question arises whether it is possible to relate to different reference frames systems of coordinates which differ one from another by a purely spatial rotation (in the simplest case when both systems are Cartesian). Such a problem becomes even more obvious when we pass from a Cartesian to polar system of coordinates.

Let us call to mind the Newtonian mechanics [see, *e.g.*, Newton (1962), Khaikin (1947), H. Goldstein (1965), Schmutzer (1989), as well as Schmutzer and Schütz (1983)], and think over, in which cases there appear, for example, the inertial forces (we do not intend to take part in scholastic discussions over the reality of those). What is clear is that such forces are originated by the state of motion of the reference frame. But if we consider two reference frames at rest, one with respect to another,— could any difference between them be physically perceivable in the course of some experiments, at least in principle? If this would be the case, it would be possible to experimentally determine, what kind of coordinate frame (*e.g.*, Cartesian or polar one) is used by the observer, or, at least, which of the relatively resting bases corresponds to the latter. It is however clear that experiments are performed irrespective to the choice of purely spatial coordinates (the corresponding system of coordinates is chosen only at the stage of mathematical description of physical effects, and the experimental devices — even if they are used for measurement of lengths or angles — have no direct relation to this choice). The same also applies to the choice of a purely spatial basis. Thus transcriptions of all expressions and quantities supposed to describe physical effects, have to be performed invariantly with respect to the choice of spatial coordinates or spatial basis. This demand is not as mild as it may seem to be at the first sight, and it yields a set of geometrical identities to be imposed on the corresponding congruences.

Hence we see that a transition between different reference frames should basically reflect their relative motion. The simplest special case of such a motion is that which relates two non-identical inertial reference frames both in Newtonian and relativistic mechanics.

It is also obviously possible to an observer or a system of observers to move together with a frame of reference, locally or in a space-time region (globally). When an object moves together with a reference frame, it is geometrically identified with the latter, so that if this is a spatially extended object, the world lines of its mass points form a congruence. Such a congruence presents a complete characterization of the reference frame, and with any reference frame a conceptual object (thus a test one) of the above type can be associated which is called the body of reference. Since it models a set of observers and/or their measuring devices, but not photons, the reference frame congruence should be time-like. A transition to another reference frame means merely a change of such a congruence, with the corresponding and automatic change of the test reference body. In order to avoid mathematical difficulties, one necessarily has to use congruences, since otherwise there would appear a danger of either mutual intersections of the world lines (which mathematically manifests itself as a non-physical singularity), or in the space-time there would arise bald spots whose boundaries will represent singularities too.

The simplest way to describe a reference frame is to identify the congruence of the corresponding reference body mass-points with a congruence of the time coordinate lines. The latter ones should be, of course, time-like, which is not always the case for the so-called time coordinate (*e.g.*, for the Schwarzschild space-time in the curvature coordinates, the t coordinate lines become null on the horizon and space-like inside it), thus we come to a restriction concerning the choice of coordinates adequate for the reference frame description of physical situations. Here a strange (from the point of view of naïve psychology) fact arises: it is easy to see that any transformation of the time coordinate with all other (spatial) coordinates fixed, can at most change the lines of the latter ones, but *not* the time coordinate lines, *i.e.* we remain with the former reference frame if determined through the time coordinate description. Such transformations are called the chronometric ones, and the invariance of a reference frame with respect to them gives a basis for Zel'manov's formalism of chronometric invariants [Zel'manov (1956, 1959); see also Mitskievich and Zaharow (1970), Mitskievich (1969)], first developed for cosmological applications.

Another formalism, that of kinemetric invariants, was proposed and developed by Ehlers (1961, 1993) and Zel'manov (1973). It is based on a description of reference frames with help of families of space-like hypersurfaces in the space-time, thus the time-like vector field normal to the hypersurfaces, exists. This vector field τ then describes non-rotating reference frames, clearly a special case in comparison to the chronometric invariants approach. We confine ourselves here to the corresponding reference frame since, as well as the formalism of chronometric invariants, this is in fact a special choice of the system of coordinates in the monad formalism extensively discussed below. Zel'manov did obviously never know about the outstanding work of Ehlers[1]. The Ehlers–Zel'manov monad formalism is accepted as the base in formulation of the reference frames theory in this book. The main differences we have to introduce here, are the use of τ (and not u) in the notation for the monad [Ehlers used a co-moving frame which is most natural in a study of cosmological problems where u is the four-velocity of matter, but in our more general approach another time-like unit vector field (τ) has to be applied, that of the four-velocity of the body of reference], and another treatment of the time derivative (the Lie derivative with respect to the vector field τ which is more natural as a comparison of objects along the worldlines of points of the body of reference).

See good reviews of all three formalisms in Zel'manov and Agakov (1989), Massa (1974 a,b,c) (an independent treatment of the problem), and partially in Shteingrad (1974), Mitskievich (1972), Ivanitskaya, Mitskievich and Vladimirov (1985, 1986). It should be noted that the formalism of chronometric invariants, with some modifications (which bring it near to the monad formalism), was independently developed by Cattaneo (1958, 1959, 1961, 1962) and Schmutzer (1968), and some applications to the study of physical effects are given in Schmutzer and Plebanski (1977) and Dehnen (1962). In its turn, the formalism of kinemetric invariants represents a refinement of the earlier formalism of Arnowitt, Deser and Misner (ADM) [see Misner, Thorne and Wheeler (1973), as well as Mitskievich and Nesterov (1981), Mitskievich, Yefremov and Nesterov (1985)]. This ADM formalism proved to be of importance for realizing the canonical formalism and the quantization procedure in gravitation theory. For applications of an approach similar to the monad formalism, to astrophysical effects, see (Thorne and Macdonald 1982), (Macdonald and Thorne 1982), (Thorne, Price and Macdonald 1986). An extensive review of methods of description of reference frames see in a monograph by Vladimirov (1982) who himself has made a substantial contribution to this field, and also in (Jantzen, Carini and Bini 1992).

During many years the principal role in description of reference frames in general relativity, was played however by the tetrad formalism which influenced much the very style of thinking. In the case of an orthonormalized tetrad, we are dealing not only with a time-like congruence which corresponds to four of the sixteen components of the tetrad (the field of tangent vector to the congruence), but with three (spatial) congruences more. One does not usually care much about the invariance of expressions which have a physical meaning, although this obviously contradicts the conclusions drawn above. If we take such a care, we come to the monad formalism with auxiliary use of an arbitrary triad in the subspace orthogonal to the monad which does not practically change our further conclusions. The tetrad formalism was considered by many authors; for an orientation in the corresponding literature, see Ivanitskaya (1969), Mitskievich (1969) and, in connection with gravitational effects, Ivanitskaya (1979).

Tetrad bases are not always directly applicable to description of reference frames, *e.g.* in the case of a Newman–Penrose basis [see (Penrose and Rindler 1984a, 1984b) where also the use of spinors in constructing bases is considered]. In some calculations, it proved effective to combine together the Dirac matrices and Cartan forms when gravitational fields

---

[1] It is remarkable that both authors were primarily interested in cosmological applications.

and reference frames are studied (Mitskievich 1975). General methods of the tetrad formalism function well in the coordinates-free approach to tensor calculus, in particular in the Cartan forms formalism (see section 1.2).

We come thus to a conclusion that it is namely the monad formalism which is most closely connected with description of frames of reference, and this formalism is presented in the next chapter. However, a question is still worthwhile: is it really always indispensable to use explicitly some definition of a specific reference frame when observable physical effects are calculated from the point of view of a non-inertial observer? And in what classes of problems of theoretical physics, and at which step of calculation procedure, the formalisms of description of reference frames may be of practical use?[2]

When local effects are studied, this means that the observer may be idealized as a single point-like test mass while receiving all outside information through signals propagating with the fundamental velocity (electromagnetic and other signals), as well as detecting all other physical and geometric factors which are perceivable on his/her world line only. Then this world line will be the only essential representative of the state of motion of the observer in the calculations, so that any extension of the reference frame to the outside would be absolutely unnecessary. Moreover, if the magnitude of some predicted physical effect would to any extent depend on the concrete way of this extension, this effect could not be considered as a correctly determined physical one, since there should be an invariance of all observables (for our local observer) with respect to an arbitrary choice of the reference frame outside the observer's world line, if only this line would, with a proper smoothness, belong to the congruence which globally describes the pertinent reference frame. It is obvious that there exists infinite and continuous multitude of such reference frames mutually coinciding on a single world line. An example of calculation of a physical effect when no extension of the reference frame outside a single world line of the observer should be considered, is the case studied in section 3.3, a united description of the gravitational red shift effect and the Doppler effect. These two effects are inseparable when the space-time is non-stationary, and their distinction is a mere convention. This conventionality manifests itself in the mentioned case in the following two interpretations of the effect. When both time-like world lines (those of the emitter of radiation and of its detector) are considered to belong to the same congruence of a reference frame, both emitter and detector are at rest by definition, and the effect has to be interpreted as a purely "gravitational" one (at least, it is then due to expansion or shrinking of the three-dimensional subspace of the reference frame). The distance between the two objects is then continually changing; in a similar situation in cosmology, the non-stationarity of space-time completely prevents domination of any of the alternative interpretations. Another approach consists of inclusion in the reference frame congruence of the observer's (detector's) world line only; then the radiation source will be in motion with respect to the frame, so that the effect would be (partly) due to the Doppler shift. An interpretation of this effect as a completely Doppler one, is also possible if the space-time is stationary; the reference frame congruence is then the Killing

---

[2]Many physicists, mainly not working in the relativity theory, share the opinion that the reference frames description is merely an approximate and non-covariant approach. But this is a sheer misapprehension.

one, although it fails to be time-like in the ergosphere. But outside the latter, one has to consider the effect as being due to motions of both emitter and detector of radiation. As to the time-like Killing congruence, if it is rotating, one usually connects it with the dragging phenomenon. Such an objectively determined reference frame is discussed in section 3.4.

There still exists, however, a wide class of other problems in which employment of a concrete global reference frame is perfectly acceptable and desirable, while the predicted in this way effects do depend on its specific choice. These are the problems in which the interacting physical factors possess spatial extension, *e.g.* the case of charged perfect fluid moving in electromagnetic and gravitational fields. In section 4.4[3] we consider such an exact self-consistent problem which moreover yields a strange (at first sight) result that for the state of motion under consideration, being consistent with behavior of fields, a non-test charged perfect fluid produces in its co-moving reference frame only a magnetic, but no electric field whatsoever. Hence its particles, being undoubtedly electrically charged, do not mutually interact electromagnetically. We show also that there is no paradox, but merely an encounter with unusual (for the prevailing inertial' mode of thinking) specific feature of electromagnetic field in non-inertial, in particular, rotating reference frames. We shall discuss there a similarity of this situation with that in classical mechanics with its centrifugal and Coriolis forces. The concept of a reference frame and employment of formalisms of its description is certainly fruitful in the relativistic cosmology where the continuous medium filling the universe, realizes a privileged (in a certain sense, *really existing*) reference frame. The fact of a non-test nature of this medium does not mean that the reference frame perturbs in this case geometry, this is merely a specific — co-moving — frame, which is objectively singled out of all other frames by the very statement of the problem.

We do not (and objectively cannot) give here an unambiguous and precise answer to the question what problems demand application of the concept of global reference frame, and for what problems this is unnecessary. Instead we confine ourselves to certain examples, since it would be a scholastic task to foresee every special case of application of reference frames for all the future. As to introduction of this concept at an initial stage of every calculation, this clearly cannot be practical, since symmetry of a problem may more easily be related to a choice of system of coordinates and not with a reference frame, so that the adequate choice of coordinates is more crucial for the most effective treatment of problems (*e.g.*, when Einstein–Maxwell fields are derived). When however one turns to interpretation of the obtained solutions, introduction of reference frames becomes quite opportune, and it may play a decisive role. These considerations bring us to a better understanding of the fact that the first of all complete and practical methods of description of reference frames was Zel'manov's formalism of chronometric invariants (the quotient system of coordinates co-moving with a frame of reference), and this was primarily done in problems of relativistic cosmology. In the same context, certain problems connected with the Noether theorem were also considered by Mitskievich and Mujica (1967), and Mitskievich and Ribeiro Teodoro (1969).

The concept of reference frame should be probably essential also at starting points of

---

[3] See also section 4.5 where a purely kinematic magnetic monopole distribution is considered.

development of new theoretical ideas. As an example the idea of zero-point radiation may be mentioned which seems to be still at a relatively early stage of its development. We only recall that the zero-point radiation possesses such a spectrum that under Lorentz transformation, it retains its form (while a black-body radiation would alter its temperature), but a transition to a uniformly accelerated frame yields in addition to the former zero-point radiation, also a black-body thermal one (Boyer 1980). The reader should not be perplexed by the fact that the integral energy of the zero-point radiation diverges (it is usually treated in the spirit of renormalization ideas). These items are closely related to the concept of the Rindler vacuum, and they accord with the quantum theory of black holes and the black hole evaporation (Thorne, Price and Macdonald 1986). In quantum physics the concept of reference frame has however not yet found its adequate realization. At the same time, its possible application areas are quite far-reaching: this is the theory of quantum fields itself (Gorbatsievich 1985) as well as motivation and fundamentals of the field quantization [the ADM formalism to be once more mentioned (Misner, Thorne and Wheeler 1973), alongside of a new development in this area — Ashtekar's canonical formalism (Ashtekar 1988), some early applications of which are sketched out in the concluding chapter of this book].

As it was motivated above, we have chosen the monad approach for description of frames of reference, where monad means a fixed unit time-like vector field which is naturally interpreted as a field of four-velocities of local observers equipped with all necessary measurement devices. In the space-time region where such a reference frame is realized, we may therefore consider a congruence of integral curves of this vector field, *i.e.* of the worldlines of particles forming a reference body. Every idealized local observer represents such a test particle. The classical (non-quantum) physics approach admits that acts of measurement do not influence the objects and processes which are subjected to the measurement, in full conformity with our treatment of local observers as test particles, — in other words, in conformity with the identification of a reference frame and a time-like congruence which, of course, should be geometrically admissible in the space-time region under consideration. Then a transition to another reference frame should be equivalent to a new choice of such an admissible congruence, without any change of the four-geometry of the world itself.

Rodichev (1972, 1974) has undertaken an attempt to overpass the limits of this approach to reference frames, but we consider his attempt as a premature and technically unacceptable one, though it is worth making an acquaintance with his concepts which were expressed with much inspiration. These concepts might become useful for far-reaching generalizations in the future, such as those connected with quantum gravity. As a matter of fact, in quantum theory it is useless to try to introduce test particles in the classical meaning, and moreover, the local observers cannot be considered there without perturbations of the four-geometry, as well as of all involved physical objects. Thus one has to consider a radical revision of the concept of reference frame when passing to quantum physics. Some ideas of such a generalization were outlined by Fock (1971) [see also (Schmutzer 1975)], and some (rather obscure) remarks were made by Brillouin (1970), but probably the most powerful insight can be found in the early paper by Bohr and Rosenfeld (1933) [see also Regge (1958) and Anderson (1958a,b; 1959)]. We intend here, however, to confine ourselves to

2.2 Reference Frames Algebra

The first steps in the development of the monad formalism were made already in the forties and fifties by Eckart (1940) and Lief (1951) on the basis of hydrodynamical analogies. In our general relativistic approach, this means that in the region under consideration, a timelike four-vector monad[4] field τ is introduced in addition to the four-dimensional metric g, with a normalization condition $\tau \cdot \tau = +1$. Then a symmetric tensor,

$$b = g - \tau \otimes \tau, \qquad (2.2.1)$$

will be automatically the projector onto the local subspace orthogonal to τ:

$$\left. \begin{array}{ll} b(\tau, \cdot) = 0, & b^\lambda_\lambda = 3, \\ \det b_{\mu\nu} = 0, & b_{\mu\nu} b^{\lambda\nu} = b^\lambda_\mu. \end{array} \right\} . \qquad (2.2.2)$$

As we shall see now, the tensor b plays simultaneously the role of a metric in this subspace (the three-dimensional physical space of the reference frame). The three-space is then non-holonomic in general (namely, when the τ-congruence is rotating).

Consider now an arbitrary four-(co)vector q; its projection onto the monad is a scalar,

$$\overset{(\tau)}{q} := q \cdot \tau, \qquad (2.2.3)$$

and the projection onto the three-space of the reference frame, a four-dimensional (co)vector,

$$\overset{(3)}{q} := b(q, \cdot), \qquad (2.2.4)$$

which is by a definition orthogonal to the monad and not changing by a repeated projection. Hence,

$$q = \overset{(\tau)}{q} \tau + \overset{(3)}{q} . \qquad (2.2.5)$$

Applying the same procedures to another (co)vector, p, one can write now

$$p \cdot q \equiv g(p, q) = \overset{(\tau)(\tau)}{p \; q} + \overset{(3)}{p} \cdot \overset{(3)}{q} =: \overset{(\tau)(\tau)}{p \; q} - \overset{(3)}{p} \bullet \overset{(3)}{q}, \qquad (2.2.6)$$

where we introduced the notation $\bullet$ for a three-dimensional scalar product relative to the reference frame τ. The sign before this three-scalar product is chosen according to signature of the tensor b, which can be written symbolically as

$$\operatorname{sign} b = (0, -, -, -), \qquad (2.2.7)$$

---

[4]In fact, this a kind of four-velocity field which is in one-to-one correspondence with the reference frame being thus introduced. We dare to reiterate that the notation τ is chosen here since this four-velocity in general will not coincide with the four-velocity u of a pointlike particle or (locally) a perfect fluid, if these are involved in the material system under consideration.

it guarantees non-negativity of such a three-scalar square of any vector. Now observe that

$$p \bullet q = -b_{\mu\nu} p^\mu q^\nu \equiv *[(\tau \wedge p) \wedge *(\tau \wedge q)], \quad (2.2.8)$$

where the (co)vectors p and q are considered to be arbitrary (four-dimensional), and their projection onto the physical three-space of the reference frame is provided automatically by the projecting properties of the three-metric.

Let us apply this simple result to the squared interval using notations dt and $d\overset{(3)}{x}$ for the elements of physical time and three-spatial displacement (relative to the reference frame under consideration) correspondingly, according to the definitions of projections, (2.2.3) and (2.2.4). These elements are, of course, not total differentials (*i.e.*, they are non-holonomic). Thus

$$ds^2 = dt^2 - dl^2, \quad dl^2 := -d\overset{(3)}{x} \cdot d\overset{(3)}{x} \equiv -b(dx, dx), \quad (2.2.9)$$

and for propagation of a light pulse (signal),

$$ds^2 = (1 - v^2) dt^2 = 0 \implies v = 1, \quad (2.2.10)$$

where $v = dl/dt$ is the absolute value of the three-velocity. This result is a standard assertion of the universality of the speed of light (but now in an arbitrary reference frame and in any gravitational fields). This conclusion does not contradict the experimental facts of the time-delay of electromagnetic signals when propagated near large masses (*e.g.*, that of the Sun), since in such experiments, the distances were calculated in the same sense as in our approach, but the physical time intervals were measured by a clock on the Earth, while in the expression (2.2.10), the local clocks are supposed to be used which should be localized continuously along the path of the signal (this means, of course, merely the corresponding re-calculation, as it was the case for the distances).

It is clear that the four-velocity of any object can be easily expressed through the monad vector and the three-velocity of this object,

$$u = \frac{dt}{ds}(\tau + v), \quad v = b(\frac{dx}{dt}, \cdot), \quad (2.2.11)$$

while

$$\frac{dt}{ds} = \overset{(\tau)}{u} = (1 - v^2)^{-1/2}. \quad (2.2.12)$$

Multi-index quantities (components of tensors) are to be projected in each of their indices onto either the monad, or the three-space orthogonal to the latter, so that the whole family of objects of various tensor ranks arises (some examples can be found below).

Let us introduce the three-spatial (and hence three-index) axial Levi-Cività tensor,

$$e_{\lambda\mu\nu} := \tau^\kappa E_{\kappa\lambda\mu\nu}. \quad (2.2.13)$$

Practically, we shall not use this objects, but its existence is obvious, and considering it helps to understand more profoundly some expressions below. Similarly to (1.2.3) we introduce also the skew tensor b_b^a, with antisymmetrization in all individual indices forming each of the collective ones. Then

$$\left.\begin{array}{ll} e_{\lambda\mu\nu}e^{\alpha\beta\gamma} = -0!\,3!\,b_{\lambda\mu\nu}^{\alpha\beta\gamma}; & e_{\alpha\mu\nu}e^{\alpha\beta\gamma} = -1!\,2!\,b_{\mu\nu}^{\beta\gamma}; \\ e_{\alpha\beta\nu}e^{\alpha\beta\gamma} = -2!\,1!\,b_{\nu}^{\gamma}; & e_{\alpha\beta\gamma}e^{\alpha\beta\gamma} = -3!\,0! \end{array}\right\}. \qquad (2.2.14)$$

As it was the case in the flat three-dimensional world for the Levi-Cività symbol, the axial tensor (2.2.13) makes it possible to relate rank two skew tensors to the corresponding vectors which also belong to the three-space of the reference frame. This fact becomes, as usually, the basis for definition of the vector product [and, later, the curl operation, (2.4.3)]. Its symbolic expression reads

$$p \times q := *(p \wedge \tau \wedge q). \qquad (2.2.15)$$

As in the scalar product (2.2.8), one may take here simply the *four*-vectors p and q without specially projecting them onto the three-space of the reference frame, while the resulting vector product will depend on the properly projected vectors, and it belongs itself to this three-space. The corresponding expression for a mixed triple product is

$$n \bullet (p \times q) = *(n \wedge p \wedge q \wedge \tau). \qquad (2.2.16)$$

It is convenient to introduce 1-form quantities

$$b^{(\alpha)} = \theta^{(\alpha)} - \tau^{(\alpha)}\tau, \quad b_{(\alpha)} = X_{(\alpha)} - \tau_{(\alpha)}\tau \qquad (2.2.17)$$

(the coordinated basis could be also used). Now the three-metric itself arises as a three-scalar product,

$$*(b^{(\alpha)} \wedge *b^{(\beta)}) \equiv *(b^{(\beta)} \wedge *b^{(\alpha)}) = b^{(\alpha)} \bullet b^{(\beta)} = -b^{(\alpha)(\beta)}, \qquad (2.2.18)$$

while we have, naturally,

$$*(b^{(\alpha)} \wedge *\theta^{(\beta)}) \equiv *(\theta^{(\alpha)} \wedge *b^{(\beta)}) = -b^{(\alpha)(\beta)}. \qquad (2.2.19)$$

Other important formulae involving τ and $b^{(\alpha)}$ are

$$dx^{\alpha\beta} = \tau^{\alpha}\tau \wedge b^{\beta} - \tau^{\beta}\tau \wedge b^{\alpha} + b^{\alpha} \wedge b^{\beta} \qquad (2.2.20)$$

and

$$dx^{\alpha\beta} = \tau^{\alpha}\tau \wedge b^{\beta} - \tau^{\beta}\tau \wedge b^{\alpha} - e^{\alpha\beta}{}_{\gamma} * (\tau \wedge b^{\gamma}). \qquad (2.2.21)$$

Transition from the first to the second relation means that there holds a characteristic triad identity

$$b^{\alpha} \wedge b^{\beta} = b^{\alpha} \times b^{\beta} = -e^{\alpha\beta}{}_{\gamma} * (\tau \wedge b^{\gamma}) \qquad (2.2.22)$$

which can be easily checked: the four quantities $b^{(\alpha)}$ correspond to three degrees of freedom only.

As we have seen, the monad formalism was first independently introduced by Ehlers (1961) and Zel'manov (1976). This monad description of reference frames is closely connected with the formalism of chronometric invariants [which belongs also to Zel'manov (1956)] by a special choice of the system of coordinates.[5] It is important that both approaches do not involve any restrictions upon the choice of a reference frame, except for the standard limitations due to geometry and physics in general, but the monad formalism allows in addition also an arbitrary choice of systems of coordinates which are then by no means related to a reference frame.

2.3 Geometry of Congruences. Acceleration, Rotation, Expansion and Shear of a Reference Frame

As we have seen above, when we discussed in section 2.1 the introduction of reference frames formulation, the crucial rôle in their description has to be played in a study of congruences by the world lines of particles forming bodies of reference, *i.e.* the physical time lines for the chosen reference frame. The congruence concept is essential because for the sake of regularity of the mathematical description of the frame, these lines have not to mutually intersect, and they must cover completely the space-time region under consideration, so that at every world point one has to find one and only one line passing through it. Exactly the same approach is used for description of a regular continuous media, *i.e.* in hydrodynamics of a compressible fluid. Therefore it is worth using here the standard formalism of hydrodynamics [relativistic: Lichnerowicz (1955), Taub (1978), non-relativistic: S. Goldstein (1976)]. In a line with these ideas we introduce below the concepts of acceleration, rotation and rate-of-strain for a frame of reference.

Differential operations on a manifold can (and for our purposes have to) be projected onto the physical time direction of a frame of reference and its local three-space with help of the projectors τ and b, as it was the case for tensor quantities, in particular, exterior forms. It is however important to keep in mind that when quantities lying in the three-space of a reference frame are being differentiated, new terms arise which are non-orthogonal to the monad field. Hence in order to obtain after such a differentiation purely three-spatial quantities, one has to perform projections not only in the differentiation indices, but also in those which stem from the very quantities being differentiated: in general, neither the three-

---

[5]Zel'manov's formalism of chronometric invariants is in fact the monad formalism written in coordinates co-moving with the monad, *i.e.* such that $\tau^\mu = \delta_0^\mu / \sqrt{g_{00}}$. This means that in this case the monad congruence coincides with the time coordinate lines. This coincidence does not change under arbitrary transformations of the time coordinate (so that the latter may arbitrarily mix with the spatial coordinates), but the new spatial coordinates should be (any) functions of the old *spatial* coordinates only. This special choice of τ substituted into all algebraic and differential relations of the monad formalism, yields a fairly complete structure of Zel'manov's formalism of chronometric invariants. The latter may seem to be intuitively more manageable than the monad formalism is, but in the reality relations involving chronometric invariants are more complicated; moreover, in the monad formalism one may attain additional simplifications by choosing special systems of coordinates which are by no means constrained by an already fixed reference frame. This is why we shall not consider below the formalism of chronometric invariants in more detail.

projector, nor the monad vector are constant relative to the differentiation operations, if the latter ones are not reformulated specially to this end. We mean here the operations with the properties of the type of (1.2.44), which are to be used for this new differentiation. Another operation which is evidently destined for taking a derivative with respect to the physical time of a reference frame, is the Lie derivative with respect to τ; but it does not commute either with the covariant components of τ, or with the three-projector (both covariant or contravariant). Moreover, the Lie derivative, to a greater extent than the nabla operator, depends on the choice of norm of the vector with respect to which the differentiation is performed (a scalar function cannot be brought out of this vector since it enters the Lie derivative under the differentiation sign, see (1.2.20)). We have made use of this "deficiency" when discussing hypersurface geometry in (Mitskievich, Yefremov and Nesterov 1985) — the integrating factor method works here in the case of normal congruences only.

The formalism we treat below, is well fitted not only for the reference frames description, but also, to a full extent, in studies of relativistic hydrodynamics. In an analogy with four-dimensional hydrodynamics, we may write the covariant gradient of monad vector as

$$\tau_{\mu;\nu} = \tau_\nu G_\mu + A_{\nu\mu} + D_{\nu\mu}, \qquad (2.3.1)$$

where G is the *acceleration (co)vector* of reference frame; A the *angular velocity tensor* or *rotation* of the frame; and D, the *rate-of-strain tensor* (the sum $A + D$ represents the purely spatial part of the gradient, which was further separated into symmetric and skew parts). Let us consider now these quantities more closely.

Acceleration is the most obvious of these constructions. It is clear that it should be equal to $\nabla_\tau \tau$; a substitution of the expression (2.3.1) brings this to

$$\nabla_\tau \tau = \tau^\lambda{}_{;\mu} \tau^\mu \partial_\lambda = G. \qquad (2.3.2)$$

Since the norm of τ is equal to unity by a definition, $\tau^\mu \tau_{\mu;\nu} \equiv 0$, and all other terms lie completely in the three-space of the reference frame. From the definition of Lie derivative and (2.3.1), we know that[6]

$$\pounds_\tau \tau_\mu = \tau_{\mu;\nu} \tau^\nu + \tau^\nu{}_{;\mu} \tau_\nu = G_\mu \qquad (2.3.3)$$

which yields

$$\pounds_\tau b_{\mu\nu} = \pounds_\tau g_{\mu\nu} - \tau_\mu G_\nu - \tau_\nu G_\mu = 2D_{\mu\nu} \qquad (2.3.4)$$

[see also (2.3.1)]. This conforms to the understanding of the rate-of-strain tensor as a measure of how the spatial scales are evolving, these being determined by the three-metric b. Finally, rotation of the τ-congruence is determined as a covector

$$\omega := *(\tau \wedge d\tau) = 2*(\tau \wedge A), \quad A := \frac{1}{2} A_{\mu\nu} dx^{\mu\nu}. \qquad (2.3.5)$$

---

[6] One should not write here $\pounds_\tau \tau$ since $\pounds_\tau \tau^\mu \equiv 0$, and there is no context which could help in discriminating between co- and contravariant vector τ.

(from here one-to-one correspondence is obvious between the bivector A and axial (co)vector ω, both bearing the same information about rotation of the reference frame). Hence

$$A = -\frac{1}{2} * (\tau \wedge \omega), \quad G = - * (\tau \wedge *d\tau). \tag{2.3.6}$$

The rate-of-strain tensor (2.3.4) splits naturally into its trace (the scalar of expansion, or dilatation)

$$\Theta := \frac{1}{2}\tau^\alpha{}_{;\alpha} = \frac{1}{2}D^\alpha_\alpha \equiv \frac{1}{2}D_{\mu\nu}b^{\mu\nu} \tag{2.3.7}$$

and the traceless part (the shear tensor),

$$\sigma_{\mu\nu} := D_{\mu\nu} - \frac{2}{3}\Theta b_{\mu\nu}, \tag{2.3.8}$$

whose square

$$\sigma^2 \equiv \sigma_{\mu\nu}\sigma^{\mu\nu} \equiv \sigma_{\mu\nu}D^{\mu\nu} = D_{\mu\nu}D^{\mu\nu} - \frac{4}{3}\Theta^2 \tag{2.3.9}$$

invariantly characterizes the presence of shear by evolution of the three-space of a reference frame.[7]

Consider now exterior differential of the monad covector,

$$d\tau = \tau \wedge G + 2A, \tag{2.3.10}$$

which can be rewritten as

$$d\tau = -G \wedge \tau + *(\omega \wedge \tau) \tag{2.3.11}$$

[cf. (3.1.18) for the electromagnetic field case], moreover,

$$\delta\tau = \tau^\alpha{}_{;\alpha} = 2\Theta. \tag{2.3.12}$$

The relation (2.3.11) is a part of Cartan's first structure equations adapted to a monad basis which can be supplemented (without leaving the coordinated basis approach) by quantities similar to (2.2.17), a set of four covectors,

$$b^\mu = dx^\mu - \tau^\mu \tau, \tag{2.3.13}$$

and four vectors,

$$b_\mu = \partial_\mu - \tau_\mu \tau. \tag{2.3.14}$$

One has to keep in mind that the free index μ appearing in these expressions, simply enumerates these covectors and vectors, and does not contribute to the tensor properties of the involved quantities from the point of view of the coordinate-free operations (∇ or

---

[7] These concepts are important also in theory of null (light-like) congruences often used in the classification of gravitational fields by principal null directions (the Petrov types) and generation of exact Einstein–Maxwell solutions [see Kramer et al. (1980) and the 2nd edition, Stephani et al. (2003)]. When a null complex (Newman–Penrose) tetrad is considered, some of the above relations change their form.

exterior differentiation), thus it spoils the properties of the components of these derivatives, so that Christoffel symbols enter them, *e.g.*

$$db^\mu = -2\tau^\mu A + \left[\tau^\mu G_\nu - (D_\nu{}^\mu + A_\nu{}^\mu) + \Gamma^\mu_{\lambda\nu}\tau^\lambda\right] b^\nu \wedge \tau. \qquad (2.3.15)$$

Here these symbols play role of scalar functions (from the point of view of exterior forms formalism). One can however get rid of them by introduction of a new differential operation

$$\overset{(b)}{d} := b^\alpha \nabla_{b_\alpha} \wedge \equiv b^\alpha \nabla_{\partial_\alpha} \wedge \equiv d - \tau\nabla_\tau \wedge. \qquad (2.3.16)$$

Thus we come to one more notation:

$$\overset{(\tau)}{d} := \tau\nabla_\tau \wedge, \qquad (2.3.17)$$

consequently

$$d = \overset{(\tau)}{d} + \overset{(b)}{d}. \qquad (2.3.18)$$

Making use of the relation

$$\overset{(\tau)}{d} b^\mu = (\Gamma^\mu_{\nu\lambda}\tau^\lambda b^\nu + \tau^\mu G) \wedge \tau, \qquad (2.3.19)$$

we obtain finally

$$\overset{(b)}{d} b^\mu = -2\tau^\mu A - (D_\nu{}^\mu + A_\nu{}^\mu)b^\nu \wedge \tau. \qquad (2.3.20)$$

Here [*cf.* (2.3.5)] $A = (1/2)A_{\mu\nu}b^\mu \wedge b^\nu$. The price to be payed for this simplification, is incompleteness of the system of structure equations (2.3.10) and (2.3.20), which however does not prevent obtaining from them the complete information about the most important quantities G, D, and A which depend only on the choice of monad. Let us simultaneously write down relations analogous to (2.3.2):

$$\nabla_{b_\mu}\tau = (D_\mu{}^\nu + A_\mu{}^\nu)b_\nu, \qquad (2.3.21)$$

$$\nabla_\tau b_\mu = -\tau_\mu G - G_\mu \tau + \Gamma^\lambda_{\mu\nu}\tau^\nu b_\lambda, \qquad (2.3.22)$$

$$\nabla_{b_\mu} b_\nu = (\Gamma^\kappa_{\nu\lambda} b_\kappa - \tau_\nu \tau^\kappa_{;\lambda} b_\kappa - \tau_{\nu;\lambda}\tau)b^\lambda_\mu. \qquad (2.3.23)$$

2.4 Differential Operations and Identities of the Monad Formalism

When a reference frame is not taken into account and only the four-dimensional nature of the universe matters, the role of curl is played by exterior derivative, and role of divergence by the operation δ (1.2.45). With help of these operations the four-dimensional de-Rhamian is then built which represents a generalization of the Laplacian (d'Alembertian) operator to the theory of differential forms. However then the operation of gradient raises a problem: it is enough to call in mind its three-dimensional definition in Cartesian coordinates of the Euclidean space, $(\text{grad }\Phi)^i = \partial_i \Phi$. Thus in the monad theory one has to use the definition

$$(\text{grad }\Phi)^\mu = -b^{\mu\nu}\partial_\nu \Phi \tag{2.4.1}$$

which already evokes doubts about the usual four-dimensional form of this operator. But, leaving this problem to the reader, let us now write down the covariant version of (2.4.1):

$$\text{grad}\Phi = -b_\mu^\nu \Phi_{,\nu} dx^\mu = -\overset{(b)}{d}\Phi. \tag{2.4.2}$$

Returning to curl, let us introduce this operator in the monad formalism. To this end it is sufficient to recall its definition in three-dimensional Euclidean geometry where curl is considered as a vector product in which the first factor is the nabla operator.[8] Here we shall base on the vector product expression (2.2.15), but the monad vector τ should not be influenced by the exterior differentiation,

$$\text{curl }\overset{(3)}{a} = *(\tau \wedge d\overset{(3)}{a}) \tag{2.4.3}$$

(of course, the sign minus in (2.4.2) is here also taken into account). In this definition, curl is acting on a covector already lying in the three-space of the reference frame. Remember that in the definition (2.2.15) it was unnecessary to consider any of the covectors taking part in the vector multiplication, to be projected onto the three-space: they are projected automatically by virtue of (2.2.15). But in the case of curl this kind of definition requires to include an additional term which should compensate the part proportional to τ of the covector being differentiated[9]. It is easy to see that then

$$\text{curl }a = *(\tau \wedge da) - \overset{(\tau)}{a}\omega, \text{ thus curl }\tau = 0, \tag{2.4.4}$$

where a is an arbitrary four-(co)vector. It might seem that in this expression the overall sign should be chosen in the opposite sense, but the expression (2.4.4) is perfectly correct due to (2.4.1). Another simple way to see this is to compare the curl components (2.4.3) brought to the contravariant level and its components in the standard Euclidean vector calculus.

Let us turn now to the three-divergence operation. The natural way to define it, is to change in δ, (1.2.45), the exterior differential d to $\overset{(b)}{d}$, (2.3.16). The latter formula gives also the connection between both operations, so that

$$\text{div }\overset{(3)}{a} := -*\overset{(b)}{d}*\overset{(3)}{a} = \delta \overset{(3)}{a} + *(\tau \wedge *\nabla_\tau \overset{(3)}{a}), \tag{2.4.5}$$

---

[8] Remember that what we really need, are not approximate, yet *exact* expressions in the local subspace $\perp \tau$ even in special relativity, not only to deal with dynamical equations, but also for intuitive interpretation of physical results, especially in non-inertial frames.

[9] It is obvious that even without this compensating procedure, the resulting covector curl *would still lie* in the three space orthogonal to the monad, but then there will appear a term $\overset{(\tau)}{a}\omega$ which is related to rotation of τ and not of a (in the four-dimensional sense). Let us try to avoid this paradoxical situation.

and finally
$$\operatorname{div} \overset{(3)}{a} := \delta \overset{(3)}{a} - G \bullet \overset{(3)}{a}. \qquad (2.4.6)$$

The extension of this three-divergence operation to covectors which have not been beforehand projected onto three-space of the reference frame, is achieved by a compensation of the redundant terms, as in the curl case. The final expression of the three-dimensional divergence operation in the monad language (the point of view of a reference frame) is

$$\operatorname{div} a := \delta a - G \bullet a - 2 \overset{(\tau)}{a} \Theta - \tau d \cdot \overset{(\tau)}{a} \qquad (2.4.7)$$

(*cf.* the last footnote). This definition yields $\operatorname{div} \tau = 0$. It is worth mentioning that when the reference frame congruence rotates (the three-space is non-holonomic), the second-order operation div curl does not vanish identically, but it yields

$$\operatorname{div} \operatorname{curl} \overset{(3)}{a} = \frac{1}{2} \omega \bullet \mathcal{L}_\tau \overset{(3)}{a}; \qquad (2.4.8)$$

similarly, $\operatorname{curl} \operatorname{grad} \Phi = -2\Phi_{,\alpha} \tau^\alpha \omega$. This can be easily checked with help of relation (2.3.6) and the definition of Lie derivative. We leave to the reader to compare this result with the identity (13.9) in Zel'manov and Agakov (1989); our next task is here to consider other important differential identities for acceleration, rotation, and rate-of-strain tensor, in particular those involving also curvature.

The first of them is a scalar identity most easily obtained when one applies operator δ to the rotation vector ω:

$$\delta \omega = \omega^\alpha{}_{;\alpha} = 2\omega \bullet G \Rightarrow \operatorname{div} \omega = \omega \bullet G. \qquad (2.4.9)$$

This identity means that in the derivatives of ω which enter $\delta \omega$, the second derivatives of monad τ cancel out. The same is true for the magnetic field vector B which is a part of dA (A being the electromagnetic four-potential covector), and the second derivatives of A, *cf.* (4.3.11).

Consider now curl applied to the acceleration of a reference frame. The resulting covector should lie in the three-space, hence all terms proportional to τ must vanish. But first we have

$$\operatorname{curl} G = - * [G \wedge *(\tau \wedge \omega)] - *d(\tau \wedge G), \qquad (2.4.10)$$

where the first term is collinear to τ, so that it will be exactly compensated, while the last term should be rearranged in such a way that it would not lead to a vicious circle. To this end we use here relations $d(\tau \wedge G) = -2dA$ [see (2.3.10)],

$$\mathcal{L}_\tau \omega = (\omega_{\beta;\alpha} \tau^\alpha + \omega_\alpha \tau^\alpha{}_{;\beta}) dx^\beta$$

and

$$\tau_{\alpha;\beta} + \tau_{\beta;\alpha} = G_\alpha \tau_\beta + G_\beta \tau_\alpha + 2D_{\alpha\beta},$$

from where

$$\operatorname{curl} G = -(\omega^\alpha{}_{;\alpha} - 2G \bullet \omega)\tau + \pounds_\tau \omega - 2\omega^\alpha D_{\alpha\beta} dx^\beta + 2\Theta\omega. \qquad (2.4.11)$$

Here the first right-hand side term (the parentheses) vanishes by virtue of (2.4.9), as we have expected.

In order to find identities involving curvature, we already introduced alongside with the FW-curvature (1.2.40), also τ-curvature making use of the operator (1.2.42) which has remarkably better properties that those of (1.2.39). Thus

$$\overset{[\tau]}{\mathbb{R}}(w,v) := \nabla^\tau_w \nabla^\tau_v - \nabla^\tau_v \nabla^\tau_w - \nabla^\tau_{[w,v]}. \qquad (2.4.12)$$

This τ-curvature has two pairs of skew indices, but there is no analogue of Ricci identities for it, so that one has to take a linear combination of its components in order to obtain a three-space tensor with properties of the standard Riemann–Christoffel tensor [*cf.* a different approach in Zel'manov and Agakov (1989)]:

$$r_{\alpha\beta\gamma\delta} = \frac{1}{6}(2\,\overset{[\tau]}{R}_{\alpha\beta\gamma\delta} + 2\,\overset{[\tau]}{R}_{\gamma\delta\alpha\beta} + \overset{[\tau]}{R}_{\gamma\beta\alpha\delta} + \overset{[\tau]}{R}_{\alpha\delta\gamma\beta} + \overset{[\tau]}{R}_{\alpha\gamma\beta\delta} + \overset{[\tau]}{R}_{\beta\delta\alpha\gamma}).$$

This combination automatically satisfies Ricci type identities since the constructing-material object (here, $\overset{[\tau]}{R}$) is skew in its first and last pairs of indices; this antisymmetry is also inherited by $r_{\alpha\beta\gamma\delta}$, thus also yielding symmetry under a collective permutation of pairs $[\alpha\beta]$ and $[\gamma\delta]$, so that $r_{\beta\gamma} := r_{\alpha\beta\gamma\delta} b^{\alpha\delta}$ (of the type of Ricci tensor) becomes symmetric.

Since the operator (2.4.12) identically annuls τ, we apply it to the basis vector b_λ closely related to the monad. In contrast to τ, we have here not a pure zero, but a non-trivial expression because of the "scalar" nature of the free index of b_λ with respect to coordinate-free operations. Hence

$$\overset{[\tau]}{\mathbb{R}}(b_\mu, b_\nu)b_\lambda = [R^\alpha{}_{\beta\gamma\delta} b^\kappa_\alpha b^\beta_\lambda b^\gamma_\mu b^\delta_\nu + (D_\mu{}^\kappa + A_\mu{}^\kappa)(D_{\nu\lambda} + A_{\nu\lambda})$$
$$-(D_{\mu\lambda} + A_{\mu\lambda})(D_\nu{}^\kappa + A_\nu{}^\kappa)]b_\kappa. \qquad (2.4.13)$$

For the same reasons, the identity connected with τ can be represented as

$$b^\mu \cdot [\mathbb{R}(b_\lambda, b_\mu)\tau] \equiv R_{\beta\gamma}\tau^\beta b^\gamma_\lambda = 2\Theta_{,\beta} b^\beta_\lambda - (D_\beta{}^\mu + A_\beta{}^\mu)_{;\alpha} b^\alpha_\mu b^\beta_\lambda - 2G^\beta A_{\beta\lambda}. \qquad (2.4.14)$$

We call identities (2.4.13) the old generalized Gauss equations and identities (2.4.14) the generalized Codazzi equations [see Mitskievich, Yefremov and Nesterov (1985); for the standard theory of hypersurfaces which is not based on the reference frame approach, see Eisenhart (1926, 1972)]; the generalization is meant in the sense that they are written in non-holonomic local three-submanifolds of rotating reference frames. These equations turn into the usual Gauss and Codazzi equations respectively when $\omega = 0$ (*i.e.*, for a non-rotating congruence we come to the usual theory of hypersurfaces). Since $\overset{[\tau]}{R}$ does not possess the standard properties of curvature tensor, it is however necessary to make the next step of generalization using $r_{\alpha\beta\gamma\delta}$. Then the new generalized Gauss equations for the three-curvature read

$$r_{\kappa\lambda\mu\nu} = R_{\alpha\beta\gamma\delta} b^\kappa_\alpha b^\beta_\lambda b^\gamma_\mu b^\delta_\nu + D_{\kappa\mu} D_{\lambda\nu} - D_{\kappa\nu} D_{\lambda\mu} + A_{\kappa\lambda} A_{\mu\nu}, \qquad (2.4.15)$$

so that for the corresponding Ricci (three-)tensor we have

$$r_{\lambda\mu} \equiv r_{\kappa\lambda\mu\nu} b^{\kappa\nu} = R_{\alpha\beta\gamma\delta} b^\beta_\lambda b^\gamma_\mu b^{\alpha\delta} + D^\nu_\mu D_{\lambda\nu} - D^\nu_\nu D_{\lambda\mu} + A^\nu{}_\lambda A_{\mu\nu}$$

$$= R_{\alpha\beta\gamma\delta} b^\beta_\lambda b^\gamma_\mu b^{\alpha\delta} + D^\nu_\mu D_{\lambda\nu} - D^\nu_\nu D_{\lambda\mu} - \frac{1}{4}(\omega_\lambda \omega_\mu + \omega \bullet \omega \, b_{\lambda\mu}) \qquad (2.4.16)$$

(here $D^\nu_\mu D_{\lambda\nu} - D^\nu_\nu D_{\lambda\mu} = \sigma^\nu_\mu \sigma_{\lambda\nu} - \frac{2}{3}\Theta\sigma_{\lambda\mu} - \frac{8}{9}\Theta^2 b_{\lambda\mu}$), where we have used simple identities[10] $A_{\kappa\mu} A_{\lambda\nu} - A_{\kappa\nu} A_{\lambda\mu} \equiv A_{\mu\nu} A_{\kappa\lambda}$ and $4A_\lambda{}^\nu A_{\mu\nu} \equiv \omega_\lambda \omega_\mu + \omega \bullet \omega \, b_{\lambda\mu}$. Moreover, one has to take into account that $r_{\lambda\mu}$ and $r_{\kappa\lambda\mu\nu}$ can be expressed one in terms of another in both directions since they belong to the (even local) three-subspace.

The identities we have just obtained play a fundamental role in the 3+1-splitting of Einstein's equations and in formulation of the Cauchy problem in general relativity; they are also important for a comparison of Einstein's and Maxwell's equations which we discuss in section 5.4. It is worth mentioning here another non-trivial fact: The definitions and identities given above imply that in the case of a normal congruence (no rotation) there exists an integrating factor, *i.e.* such a function N that

$$d(N^{-1}\tau) = 0. \qquad (2.4.17)$$

In a region with simple topological properties, the 1-form

$$\zeta = N^{-1}\tau \qquad (2.4.18)$$

is not only closed but even an exact one, so that

$$\zeta = dt \qquad (2.4.19)$$

(t is not necessarily the x^0 coordinate of the system used in calculations). If a quantity

$$\xi = N\tau \qquad (2.4.20)$$

is simultaneously introduced (which usually appears as a contravariant vector), a simple reciprocal norm is realized,

$$\zeta \cdot \xi = 1. \qquad (2.4.21)$$

Moreover, alongside with the trivial identity $\mathcal{L}_\xi \xi^\mu \equiv 0$, a new important identity arises,

$$\mathcal{L}_\xi \zeta_\mu = 0. \qquad (2.4.22)$$

---

[10]Antisymmetrization in four indices belonging to a three-subspace, vanishes automatically.

The projecting tensor (three-metric) has now the components

$$b^\alpha_\beta = \delta^\alpha_\beta - \xi^\alpha \zeta_\beta, \qquad (2.4.23)$$

and it is therefore constant relative to the Lie differentiation with respect to the vector field ξ:

$$\pounds_\xi b^\alpha_\beta \equiv 0. \qquad (2.4.24)$$

These facts are important in the canonical formulation of general relativity.

Similar ideas (although in spaces of other variables than the space-time ones) can be successfully formulated using Cartan's exterior forms, *e.g.*, in thermodynamics (Torres del Castillo 1992).

Chapter 3

Equations of Motion of Test Particles

3.1 The Electric Field Strength and Magnetic Displacement Vectors

When the geodesic equation (1.2.41), the first line, is written in form

$$u^{\mu}{}_{;\alpha}u^{\alpha} = 0, \tag{3.1.1}$$

its left-hand side is interpreted as the acceleration four vector — the absolute derivative of four-velocity with respect to the proper time of the particle (du^{μ}/ds in special relativity if Cartesian coordinates are used). For a charged test particle in an exterior electromagnetic field, the equation of motion can be written in a general covariant form as

$$mu^{\mu}{}_{;\alpha}u^{\alpha} = eF^{\mu\alpha}u_{\alpha} \tag{3.1.2}$$

[*cf.*, *e.g.*, Landau and Lifshitz (1971)]. Its comparison with the forms of geodesic equation (3.1.1) and (1.2.41), the second line, equivalent to each other, shows that the 3-form equation,

$$u \wedge *d(mu + eA) = 0, \tag{3.1.3}$$

or its dual conjugate (1-form equation),

$$*(u \wedge *d(mu + eA)) = 0, \tag{3.1.4}$$

are equivalent to (3.1.2). Here

$$A := A_{\mu}dx^{\mu} \tag{3.1.5}$$

is 1-form of the electromagnetic four-potential, which should not be confused with the reference frame rotation tensor (2-form) A (2.3.5); anyhow, we shall use more frequently the rotation vector ω [see also (2.3.5)]. 2-form of the electromagnetic field strength is then

$$F := dA = \frac{1}{2}F_{\mu\nu}dx^{\mu\nu}, \tag{3.1.6}$$

while m and e in (3.1.3) and (3.1.4) are the rest mass and electric charge of the test particle respectively, u being its four-velocity.

Thus we come to a conclusion that the complete expression for the Lorentz force in the right-hand side of (3.1.2), can be represented as

$$\mathcal{F} = \mathcal{F}_\mu dx^\mu = eF_{\mu\alpha}u^\alpha dx^\mu = e*(u \wedge *dA). \qquad (3.1.7)$$

Consider now a question: how could one approach logically to the introduction of the concept of strength of a physical field? As one of the most simple cases studied in the theory of relativity in great detail, electrodynamics represents an ideal example of formulation of such an approach. This suggests to admit that for a field acting on a corresponding test charge in some state of motion with a certain force, the field strength is such a factor that enters the expression of the force (*pro* unit charge value) and does not include characteristics of the motion of the test charge itself. Thus the field strength describes a field existing in space-time independently of the charges used for its detection and measurement. In electrodynamics, this is the very tensor F; in a more complicated case of gravitational field, one has to introduce a concept of the *relative field strength*, since in the geodesic equation there is no gravitational force term whatsoever, so that it is necessary to consider the geodesic deviation equation [see (5.1.1)]. Then the tensor rank of the relative strength becomes 4 and not 2, as it was the case for the field strength in electrodynamics.

We know however from the special relativistic electrodynamics how the four- and three-dimensional Lorentz force expressions are mutually related. This can be expressed from the point of view of the reference frame (or, monad τ) as

$$\overset{(3)}{\mathcal{F}} = \overset{(\tau)}{u} e(E + v \times B), \qquad (3.1.8)$$

while

$$\overset{(\tau)}{\mathcal{F}} = \overset{(\tau)}{u} eE \bullet v \qquad (3.1.9)$$

[*cf.* the decomposition (2.2.5) and the three-projection of a vector, (2.2.4), as well as definition of the three-velocity of a particle, (2.2.11)]. We rewrite now the first of these relations using the definition (2.2.15) of the vector product,

$$\overset{(\tau)}{u} E + *(u \wedge \tau \wedge B) = *(\tau \wedge *F) + *(v \wedge *F) - [\tau \cdot *(v \wedge *F)]\tau, \qquad (3.1.10)$$

where the right-hand side is also projected onto the three-space orthogonal to τ-congruence. This expression can also be written as

$$E + v \times B = F_{\alpha\beta}(\tau^\beta + v^\beta)b^\alpha. \qquad (3.1.11)$$

This relation holds for particles moving arbitrarily with respect to the reference frame, *i.e.* for any vector v with a norm less than unity. Then one can write immediately (Mitskievich and Kalev 1975)

$$E = F_{\alpha\beta}\tau^\beta dx^\alpha \qquad (3.1.12)$$

or equivalently
$$E = *(\tau \wedge *F) \tag{3.1.13}$$

[*cf.* (2.3.6)]. Due to arbitrariness of v, the remaining part yields

$$E_{\kappa\lambda\mu\nu}\tau^{\kappa}B^{\lambda}b^{\nu} = F_{\nu\lambda}b^{\lambda}_{\mu}b^{\nu}, \tag{3.1.14}$$

or better
$$E_{\kappa\lambda\mu\nu}\tau^{\kappa}B^{\lambda}b^{\mu} \wedge b^{\nu} = -F_{\mu\nu}b^{\mu} \wedge b^{\nu}. \tag{3.1.15}$$

Here the left-hand side is equal to $2 * (\tau \wedge B)$. But since $*(\tau \wedge *(\tau \wedge B)) \equiv -B$, we have after exterior multiplication of (3.1.15) by τ from the left and a subsequent dual conjugation,

$$B = *(\tau \wedge F), \tag{3.1.16}$$

an analogue of the expression (3.1.13) [*cf.* (2.3.5)]. Then as an analogue of (3.1.12), we have expression with the opposite sign (Mitskievich and Kalev 1975):

$$B = -F^{*}_{\alpha\beta}\tau^{\beta}dx^{\alpha}. \tag{3.1.17}$$

Thus we have obtained very simple and applicable for all reference frames and arbitrary gravitational fields definitions of the electric field strength and magnetic displacement vectors, E and B. We remind that they follow unambiguously from the standard form for the Lorentz force, *i.e.* from the electric and magnetic fields interpretation in the spirit of traditional theory. The definitions (3.1.13) and (3.1.17) yield automatically a decomposition

$$dA = F = E \wedge \tau + *(B \wedge \tau), \tag{3.1.18}$$

and vice versa [*cf.* (2.3.11), (2.3.6), (2.3.5)].

In order to illustrate the simplicity of application of these definitions to concrete problems, we now calculate the electromagnetic field invariants through the three-space vectors E and B using the contracted crafty identities (1.2.6) and (1.2.7). To this end let us multiply each of the identities by $\tau_{\lambda}\tau^{\mu}$ and make use of (3.1.12) and (3.1.17):

$$F_{\sigma\tau}F^{\sigma\tau} = 2(B \bullet B - E \bullet E), \tag{3.1.19}$$

$$F^{*}_{\sigma\tau}F^{\sigma\tau} = 4E \bullet B, \tag{3.1.20}$$

where the definition of three-scalar product $\bullet$ (2.2.6), (2.2.8) is also taken into account.

Let us build, with help of (3.1.18), self-dual quantities while first writing the expression dual conjugate to (3.1.18),

$$*F = -B \wedge \tau + *(E \wedge \tau). \tag{3.1.21}$$

This relation as well as its counterpart (3.1.18), can be more easily related to the component representation of the electric and magnetic fields (3.1.12) and (3.1.17) if one takes into account the identities (2.2.20) and (2.2.22).

We denote self-dual and anti-self-dual quantities using signs $+$ or $-$, correspondingly, over the root letter:

$$\overset{\pm}{F} := \frac{1}{2}(F \mp i * F) = \overset{\pm}{E} \wedge \tau \mp i * \left(\overset{\pm}{E} \wedge \tau\right). \tag{3.1.22}$$

It is easy to see that

$$\overset{\pm}{E} := \frac{1}{2}(E \pm iB). \tag{3.1.23}$$

The quantities $\overset{\pm}{F}$ possess the properties

$$* \overset{\pm}{F} = \pm i \overset{\pm}{F}. \tag{3.1.24}$$

Then the contracted crafty identities yield

$$\overset{\pm}{F}_{\mu\nu} \overset{\pm}{F}^{\mu\nu} = -4i \overset{\pm}{E} \bullet \overset{\pm}{E}, \quad \overset{\pm}{F}_{\mu\nu} \overset{\mp}{F}^{\mu\nu} \equiv 0, \tag{3.1.25}$$

both relations following from (1.2.6) as well as from (1.2.7) since these mutually coincide in this case due to the equality (3.1.24). We see that an (anti-)self-dual electromagnetic field is equivalent to a simultaneous existence of both electric and magnetic fields, though differing one from the other by a complex factor $(\mp i)$, so that if for the initial 2-form F the correspondence $F \Leftrightarrow (E, B)$ holds, for $\overset{\pm}{F}$ we have

$$\overset{\pm}{F} \Leftrightarrow \left(\overset{\pm}{E}, \mp i \overset{\pm}{E}\right). \tag{3.1.26}$$

3.2 Monad Description of the Motion of a Test Charged Mass in Gravitational and Electromagnetic Fields

The general covariant energy-momentum vector of a mass point is decomposed into the energy scalar and momentum three-vector in a standard way

$$p = mu = \mathcal{E}\tau + \mathcal{P}, \quad \mathcal{P} \cdot \tau = 0, \tag{3.2.1}$$

while

$$\mathcal{E} := p \cdot \tau = \overset{(\tau)}{u} m, \quad \mathcal{P} := \overset{3}{p} = \mathcal{E}v, \tag{3.2.2}$$

so that

$$p \cdot p = \mathcal{E}^2 - \mathcal{P} \bullet \mathcal{P} = m^2, \quad \mathcal{E} \geq m \geq 0. \tag{3.2.3}$$

The scalar mass of a particle, often called its rest mass, is constant for elementary particles only.[1]

---

[1] Till their decay; moreover, if the lifetime of a particle is short, the constancy of m or universality of this quantity is limited by the uncertainty relation for time and energy. In the case of a macroscopic particle or a body, one cannot demand that its rest mass should be constant (*e.g.*, for a radiating star or a body which throws away or evaporates its parts). Thus in general the idea of a fixed rest mass is alien to classical (non-quantum) physics, being accepted there only phenomenologically from observation of stable objects for which this constancy has an exclusively quantum nature.

We consider here from the viewpoint of a reference frame τ the splitting of eq. (3.1.2) which describes the motion of a test charged particle. Its non-gravitational part (3.1.6) has already met a comprehensive treatment in section 3.1 from where we shall borrow the corresponding expressions; now it remains to consider splitting of the term

$$*(u \wedge *dp) \stackrel{(\tau)}{=} u * [(\tau + v) \wedge *d(\mathcal{E}\tau + \mathcal{P})] \qquad (3.2.4)$$

in which it was suitable to pass to the energy $\mathcal{E}$ and three-momentum $\mathcal{P}$ characterizing the particle and already introduced in section 3.2.

It is easy to see that the four terms into which splits the right-hand side of (3.2.4), read

$$*(\tau \wedge *d(\mathcal{E}\tau)) = d\mathcal{E} - (\pounds_\tau \mathcal{E})\tau - \mathcal{E}G, \qquad (3.2.5)$$

$$*(v \wedge *d(\mathcal{E}\tau)) = 2\mathcal{E}\, v \times \omega - (\mathcal{E}\, v \bullet G - v \bullet d\mathcal{E})\tau, \qquad (3.2.6)$$

$$*(\tau \wedge *d\mathcal{P}) = -\frac{1}{2}\pounds_\tau \mathcal{P} \qquad (3.2.7)$$

[this is true only for a three-vector (here, $\mathcal{P}$) orthogonal to τ], and

$$*(v \wedge *d\mathcal{P}) = -(v \bullet \pounds_\tau \mathcal{P})\tau + v^\mu(\mathcal{P}_{\mu,\lambda} - \mathcal{P}_{\lambda,\mu})b^\lambda. \qquad (3.2.8)$$

Differentiation of the identity (3.2.3) for $m = $ constant, yields

$$\pounds_\tau(\mathcal{E}^2) = -\pounds_\tau(g^{\mu\nu}\mathcal{P}_\mu \mathcal{P}_\nu) = -2\mathcal{P}^\mu \pounds_\tau \mathcal{P}_\mu - \mathcal{P}_\mu \mathcal{P}_\nu \pounds_\tau g^{\mu\nu}, \qquad (3.2.9)$$

where $\pounds_\tau g^{\mu\nu} = -(2D^{\mu\nu} + \tau^\mu G^\nu + \tau^\nu G^\mu)$ and $v^\mu \pounds_\tau \mathcal{P}_\mu = -v \bullet \pounds_\tau \mathcal{P}$, hence

$$\pounds_\tau \mathcal{E} = v \bullet \pounds_\tau \mathcal{P} + \mathcal{E}D^{\mu\nu} v_\mu v_\nu. \qquad (3.2.10)$$

Note also that
$$\pounds_\tau \mathcal{E} - v \bullet d\mathcal{E} = (m/\mathcal{E})\pounds_u \mathcal{E}. \qquad (3.2.11)$$

As a result we obtain the splitted equations of motion of a particle having $m = $ constant: two equivalent forms for the scalar equation (the component along τ),

$$\pounds_\tau \mathcal{E} - v \bullet d\mathcal{E} = \mathcal{E}v_\mu v_\nu D^{\mu\nu} + v \bullet (eE - \mathcal{E}G) \qquad (3.2.12)$$

and
$$\pounds_u \mathcal{E} = (\mathcal{E}/m)[\mathcal{E}v_\mu v_\nu D^{\mu\nu} + v \bullet (eE - \mathcal{E}G)], \qquad (3.2.13)$$

and the vector equation (projection onto the three-space of the reference frame),

$$\dot{\mathcal{P}} = \pounds_\tau \mathcal{P} + \nabla_v^\tau \mathcal{P} = e(E + v \times B) + \mathcal{E}(-G + v \times \omega). \qquad (3.2.14)$$

In the last case, we do not employ the Lie derivative with respect to the four-velocity of the particle, since being applied to vectors, this derivative brings the quantity out of the three-dimensional subspace. In equation (3.2.14) the differentiation operation ∇_v^τ, (1.2.42), is used which takes the reference frame explicitly into account. The corresponding term can also be written as $\nabla_v^\tau \mathcal{P} = \mathcal{P}_{\mu;\nu} v^\nu b^\mu$. From (3.2.14) we see what concrete interpretation should have ∇_v^τ: the first term, $\mathcal{L}_\tau \mathcal{P}$, essentially is a partial derivative of $\mathcal{P}$ with respect to physical time, thus addition of $\nabla_v^\tau \mathcal{P}$ brings the sum to a direct analogue of $\frac{d\mathcal{P}}{dt}$ (naturally, with general covariant properties), like the elementary procedure used in classical mechanics, $\frac{d}{dt} = \frac{\partial}{\partial t} + v \bullet \nabla$.

In the resulting equations, a far-reaching analogy between gravitation and electromagnetism can be clearly traced in the aspect of the action of these fields on test particles. Below we shall further discuss this analogy (see chapters 4 and 5).

The equations of motion (3.2.12) to (3.2.14) of course hold also in special relativity when no gravitational field (curvature) is present; then the terms containing G and ω represent the non-inertial frame effects. The very last term in (3.2.14) is then clearly the Coriolis force; the centrifugal force is hidden in the next to the last term proportional to G, if acceleration of the reference frame is expressed through ω and radius vector in the case of a pure (rigid) rotation. When gravitation is present, a mixture of gravitational and inertial forces arises, these forces being inseparable one from the other (remember the equivalence of gravity and acceleration, usually considered at a more naïve level).

3.3 Motion of Photons, the Redshift and Doppler Effects

Some concrete physical effects can be however calculated and described in such a way that they are expressed through scalar quantities being functions of states of motion of the participating particles only, including, if necessary, quanta which move with the fundamental speed. While the massive objects mentioned above may be used (if their world lines do not intersect) to build a reference frame in which these participating objects are at rest, the massless quanta cannot of course be used in constructing such reference frames, since their world lines are null. Therefore one has not to unnecessarily universalize the trend towards description of all physical effects using reference frames. When they enter naturally into an experiment scheme or simplify essentially its description or calculations, they are surely advisable, but a mutual adjustment of setting an experiment and constructing a reference frame, would be of an appallingly low standard, if this would be done in order to merely realize the idea of universal application of global reference frames. As an example of such a situation we consider here a rigorous description of one of the most important relativistic effects, that of frequency shift of a signal, which in fact unifies the gravitational redshift and Doppler shift effects (in general, their unambiguous separation is impossible). Some examples of alternative separation cases can be easily invented, but, if the space-time under consideration is non-stationary, there exists no criterion for choosing a preferential interpretation. Therefore we shall not be captivated by such a scholastic attitude, and we shall consider here an example of use of a local reference frame, up to a numerical evaluation of the frequency shift effect. Such local reference frames should be chosen individually for every specific problem, and there exists no general formalism of their application, in contrast with global reference frames for which the monad formalism excellently works.

Schrödinger (1956) has proposed an invariant procedure for determination of the frequency shift effect (in the description of the cosmological redshift), but this procedure remained practically unnoticed until Brill (1972) has given in his short report a rigorous deduction of Schrödinger's formula using coordinates-free approach, the same as to be applied below. While giving here a deduction of the Schrödinger–Brill formula, we propose to the reader to draw by her/himself the corresponding figure, so simple the idea is and so advantageous this would be for its best understanding. Let the emitter and detector of signals have time-like world lines 1 and 2, respectively, while the signals propagate along null geodesics from the emitter to detector (thus the light cone has its apex on emitter's world line, and the generatrix intersects the detectors's world line). Then two signals sent from the emitter with an interval of the proper time between them, equal to some standard period, reach the detector being divided by another interval of the proper time (now, along detector's world line), different from the initial standard period. If the detector is provided with a frequency standard moving together with it and reproducing the standard period, now from the viewpoint of the proper time of the detector, it is possible to compare these two time intervals, *i.e* to measure the frequency shift by the transmission of the signal from emitter to detector. This is a fundamental principle of the (general as well as special) relativity theory that the period (or, equivalently, frequency) of signals expressed in terms of proper time, is the same, independent of the choice of the world line of the emitter of these signals, on which the proper time is to be measured. By the way, the same idea of relativity (and of the equivalence principle), makes it in principle impossible to separate Doppler shift and gravitational red sift effects. By definition, the redshift is expressed *via* the formula [see, *e.g.*, Peebles (1993)]

$$1 + z = \frac{\lambda_2}{\lambda_1} = \frac{ds_2}{ds_1}, \qquad (3.3.1)$$

where for convenience we introduced infinitesimal proper time intervals on the corresponding world lines, those for emitter and for observer, while ds_2 is the interval between two signals which reach the detector, having initially — at their departure moments on the world line of the emitter — some standard proper time interval ds_1 between them. It is clear that this formula simultaneously describes both usual gravitational redshift effect and the Doppler shift (due to motion of the emitter as well as the detector). In general one cannot speak about a relative velocity between detector and emitter, if the space-time is curved: there will be no unambiguous way to compare the four-velocities since the comparison would depend crucially on the choice of the transportation path, so that only in the case of the Minkowski space-time (absence of any genuine gravitational field) such an absolute comparison of motions is possible: this is why we are so much accustomed to the independent existence of the Doppler shift in special relativity. Thus our task now is to determine the connection between ds_1 and ds_2 for the same pair of signals.

First we shall complement the time-like world lines 1 and 2 by introducing some intermediary time-like lines (the concrete choice will not affect the results) which, as the previous lines 1 and 2, have not necessarily to be geodesics. Thus we have built a family of lines which may be parametrized holonomically when a parameter σ will vary along each of them. Then the tangent vector field of this family of lines is $q = \partial_\sigma$. The other family of lines will be that of null lines of signals propagating from the initial world line 1 to 2. These null lines are, of course, geodesics, with a canonical (affine) parameter λ and tangent vector field $k = \partial_\lambda$, so that

$$\nabla_k k = 0, \quad k \cdot k = 0; \qquad (3.3.2)$$

thus λ also will continuously enumerate individual lines of the time-like family. Let the parameter λ be holonomic too. The condition to have simultaneously two independent holonomic parameters (which may be used as coordinates), thus

$$\nabla_q k - \nabla_k q \equiv [q, k] = 0 \qquad (3.3.3)$$

(*cf.* the sixth axiom of covariant differentiation). Note, on the one hand, that the scalar product $q \cdot k$ is constant under transport along the null lines. In order to prove this fact, one has to differentiate $q \cdot k$ with respect to the parameter λ:

$$\partial_\lambda (q \cdot k) \equiv \nabla_k (q \cdot k) = k \cdot \nabla_k q = k \cdot \nabla_q k \equiv \frac{1}{2} \nabla_q (k \cdot k) \equiv 0. \qquad (3.3.4)$$

On the other hand, this is not the vector field q (introduced for the first family of lines) which is physically meaningful, but the four-velocity vectors u on the world lines 1 and 2, or equivalently, the proper time intervals along these lines. Therefore it is worth writing the conserved (by the null transport between the lines 1 and 2) quantity as

$$q \cdot k = (ds/d\sigma) u \cdot k. \qquad (3.3.5)$$

The simultaneously consistent holonomicity property of σ and λ means in particular that the difference $d\sigma$ is the same on both lines 1 and 2, if it is taken for the same pair of signals; then evidently

$$(u \cdot k \, ds)_1 = (u \cdot k \, ds)_2 \Rightarrow \frac{ds_2}{ds_1} = \frac{(u \cdot k)_1}{(u \cdot k)_2}. \qquad (3.3.6)$$

We have thus finally obtained the Schrödinger–Brill formula

$$z = \frac{(u \cdot k)_1}{(u \cdot k)_2} - 1 \qquad (3.3.7)$$

which holds both in special and general relativity and describes Doppler frequency shift in the Minkowski world, as well as gravitational redshift effect in the Schwarzschild spacetime, and cosmological redshift in the Friedmann world, to mention a few of the principal cases.[2]

---

[2] Both Schrödinger (1956) and Brill (1972) did not use the notation z accepted in the cosmological considerations, while in (Mitskievich and Nesterov 1991) we have taken a false definition of z from the expression (114,7) in Landau and Lifshitz (1973) (see however the correct expression in p. 475 of the same book, exercise 1). Now the "redshift factor" z is finally put in its traditional form [see Peebles (1993, 1997)].

Let us apply this general formula to three cases, the special relativistic longitudinal Doppler effect, the cosmological redshift, and a free fall of the detector onto a black hole.

In the special relativistic case, global Cartesian coordinates (here, only two of them, t and x) will be used, and we shall identify the time coordinate axis with the world line 2 (essentially, its segment CD), while the line 1 (representing the signals' emitter) will be a straight line going to the right with a positive (upward) inclination (only its segment AB will be used); see the Figure 3.3. Two different kinds of null lines of signals have to be considered, AC (before the encounter of the emitter and observer, which occurs at the world point O, the origin) and BD (after the encounter). These null lines are parallel to different generatrices (the broken lines) of the light cone with its vertex at O.

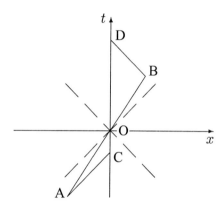

Fig. 3.3. The longitudinal Doppler effect in special relativity

Consequently, the four velocity of observer is $u^\mu = \delta_0^\mu$ while that of the emitter, $u^\mu = (1 - v^2)^{-1/2} \{1, v\}$. The null vector k is $k_- = dt - dx$ (for a photon propagating along AC) and $k_+ = dt + dx$ (propagating along BD). Thus $(u \cdot k_\mp)_1 = \sqrt{(1 \mp v)/(1 \pm v)}$ and $(u \cdot k_\mp)_2 = 1$. Hence,

$$\frac{\nu_2}{\nu_1} = \sqrt{\frac{1 \pm v}{1 \mp v}} \qquad (3.3.8)$$

(the upper sign describes the longitudinal Doppler effect before the emitter-observer encounter, and the lower sign, after the encounter). In this case, the redshift z is traditionally not used, but it can be easily found from (3.3.7).

In the second case we use the Friedmann–Robertson–Walker metric

$$ds^2 = a^2(\eta) \left[d\eta^2 - d\chi^2 - b^2(\chi)\left(d\theta^2 + \sin^2\theta d\phi^2\right)\right], \quad b(\chi) = \frac{\sin(\sqrt{\epsilon}\chi)}{\sqrt{\epsilon}}, \qquad (3.3.9)$$

$\epsilon = +1$ corresponding to the closed model, -1, to the open model, and 0, to the plane three-space (however, also in expansion which is always described by the scale factor $a(\eta)$). The simplest way is to use the coordinates co-moving with the matter of universe (to which we also relate both emitter and detector of radiation). Let the detector be at rest at the spatial origin ($\chi_2 = 0$; this is justified by homogeneity of the Friedmann universe), so that signals propagate radially. Their equation of propagation, $ds^2 = 0$, is integrated trivially: $\eta_2 = \eta_1 + \chi_1$ (it is natural to treat the moment of observation η_2 as a constant marking our epoch). Since η_1 and η_2 are cosmological times of transmission and reception of a signal, the difference between them is equal to the cosmological-coordinate "distance" of the radiation source from the origin where the observer resides (χ_1 may be also treated as an integration constant in full accordance with the use of a co-moving frame in which all the matter of the universe is at rest). Then $u = a^{-1}\partial_\eta$, $k = d\eta + d\chi$ (it is clear that u and k form geodesic fields), and (3.3.7) yields $z = a_2/a_1 - 1$. If expansion of $a(\eta_1)$ about the observation moment η_2 is performed (with the corresponding assumption that η_1 and η_2 are sufficiently close one to the other), we obtain $a_1 = a_2 - (da/d\eta)_2 \chi_1$. Since the physical distance element between the observer and radiation source is $dl = a d\chi$, this formula yields the standard Hubble "constant":

$$z \approx \left(\frac{1}{a}\frac{da}{d\eta}\right)_2 \chi_1, \quad H = \frac{dz}{dl_1} = \left[\frac{1}{a^2}\frac{da}{d\eta}\right]_2 = \left[\frac{1}{a}\frac{da}{dt}\right]_2 \qquad (3.3.10)$$

(transition to the physical time t is performed according to the relation[3] $a d\eta = dt$). In this example a system of coordinates (co-moving with the reference frame whose language is used in our discussion) was employed in which both radiation source and observer were continuously at rest, in fact, moving geodesically. However, the obtained effect is naturally treated as a Doppler shift. It is closely related to the rate of growth of distances between the galaxies (the expansion of the universe); however, in our frame the matter is everywhere at rest, but the scales are changing (the rate-of-strain tensor is non-zero for the reference frame under consideration). It is clear that there can be no physical sense in arguing into one or another interpretation of the nature of this redshift effect, the gravitational or Doppler one.

The third case is related to a Gedankenexperiment concerning a situation when over a black hole, a mother craft is hovering, and a probe dives from it with zero initial velocity into a free fall, while standard signals are continuously sent to it from the mother craft (Mitskievich and Nesterov 1991, Mitskievich 1996).[4] The problem is to find out the frequency shift of these signals received by the probe. We consider the general case of the black hole, the Kerr–Newman solution with three parameters: mass M, Kerr parameter a, and electric charge Q. In this case

$$ds^2 = \frac{\Delta - a^2 \sin^2 \vartheta}{\Sigma} dt^2 - \frac{\Sigma}{\Delta} dr^2 - \Sigma d\vartheta^2$$
$$- \frac{1}{\Sigma}\left[(r^2 + a^2)^2 - a^2 \Delta \sin^2 \vartheta\right]\sin^2 \vartheta d\phi^2 - 2a\frac{r^2 + a^2 + \Delta}{\Sigma}\sin^2 \vartheta dt d\phi, \qquad (3.3.11)$$

---

[3] This change of variables brings (3.3.9) to $ds^2 = dt^2 - a^2(t)\left[d\chi^2 - b^2(\chi)\left(d\theta^2 + \sin^2\theta d\phi^2\right)\right]$, thus these coordinates are synchronous, and along the temporal coordinate lines t coincides with s. The complete definition reads: Synchronous coordinates are those in which x^0 lines are non-rotating geodesics parameterized by the proper time along them, and x^i ($i = 1, 2, 3$) are located in global subspace orthogonal to x^0 lines.

[4] The corrected definition of z is, of course, now used.

where $\Sigma = r^2 + a^2 \cos^2 \vartheta$, $\Delta = r^2 + a^2 - 2Mr + Q^2$ (the Boyer–Lindquist coordinates). Determine first of all the tangent vectors u and k. The first integrals of geodesic equations are

$$\Delta \Sigma \dot{t} = EA,$$

$$\Sigma^2 \dot{r}^2 = E^2(r^2 + a^2) - \Delta(K + \eta r^2) =: F^2,$$

$$\Sigma^2 \dot{\theta}^2 = K - \eta a^2 \cos^2 \theta - E^2 a^2 \sin^2 \theta,$$

$$\Delta \Sigma \dot{\phi} = aE(2Mr - Q^2),$$

where $A = (r^2 + a^2)^2 - \Delta a^2 \sin^2 \theta$, E and K being the first integrals of energy and the total angular momentum [determined with help of the Killing tensor, cf. (Marck 1983)], $\eta = 0$ or 1 depending on null or time-like properties of world lines are; the speed of light and Newtonian gravitational constant are assumed to be equal to unity, and a dot denotes differentiation with respect to t. Let now $\dot{\theta} = 0$, so that the motion is radial up to such dragging effects (in ϕ direction; on the dragging effect see the next section) which cannot be excluded at all radii simultaneously.

Although the world line 1 is non-geodesic, we choose it so that it corresponds to motion of the mother ship, i.e. $u_1 = \dot{t}\,\partial_t + \dot{\phi}\,\partial_\phi$.[5] Here the expressions from above are taken for $\dot{t}$ and $\dot{\phi}$ while $\eta = 1$, and normalization of the four-velocity yields a special choice of the constant E (the Carter integral K is already fixed by the condition $\dot{\theta} = 0$). The same value of E is inherited by the probe since the world line 1 gives in fact the initial data of its motion. For the probe we have then $u_2 = \dot{t}\,\partial_t + \dot{r}\,\partial_r + \dot{\phi}\,\partial_\phi$, with the same remarks which were made with respect to u_1. Since the angle θ does not change along the world lines, the coordinate r is essential only (all quantities do not explicitly depend on t and ϕ), so that it is not necessary to determine the intersection points of the lines under consideration. In the common (initial) point of the world lines 1 and 2 (only the latter is a geodesic) both lines have a common tangent vector. As to the null geodesics, their tangent vectors are $k = \dot{t}_0 \partial_t + \dot{r}_0 \partial_r + \dot{\phi}_0 \partial_\phi$, and since all components are proportional to the same constant E_0, this can be equalized without loss of generality to the already used E. Hence $\dot{t}_0 = \dot{t}$, $\dot{\phi}_0 = \dot{\phi}$, $\dot{\theta}_0 = \dot{\theta} = 0$, and the only remaining non-trivial relation is $\Sigma^2 \dot{r}_0^2 = E^2[(r^2 + a^2)^2 - \Delta a^2 \sin^2 \theta]$. Thus $\dot{r}_0^2 = \dot{r}^2 + \Delta/\Sigma$, and $k = u_2 + (\dot{r}_0 - \dot{r})\partial_r$. Since $\dot{r}_1 = 0$, $(k \cdot u)_1 = 1$, and in general $k \cdot u = 1 + \dot{r}(\dot{r}_0 - \dot{r})g_{rr}$.[6] Then

$$(k \cdot u)_2 = 1 - \left[\frac{\Sigma \dot{r}^2}{\Delta}\left(\sqrt{1 + \frac{\Delta}{\Sigma \dot{r}^2}} - 1\right)\right]_2 \qquad (3.3.12)$$

($\dot{r}^2$ will disappear shortly, so that there is no need to substitute its expression already given above).

---

[5] The dragging-like azimuthal motion of the mother ship is mimicking the real dragging experienced by the probe in its free fall in the Kerr–Newman field.

[6] In fact, $\dot{r} = -F/\Sigma$ and $\dot{r}_0 = -EA^{1/2}/\Sigma$, where the choice of sign (minus) is determined by the direction of motion (the fall). However it is not necessary to take these relations into account in our calculations.

We see that when the probe approaches a horizon (event or Cauchy one[7]), $\Delta \to 0$, $(k \cdot u)_2 \to \frac{1}{2}$, the limiting value of the redshift is $z = 1$. It is remarkable that this redshift does not depend on anything whatsoever — on mass, Kerr parameter, charge of the black hole, the radius r_1 at which the mother craft is hovering (with a drift in the azimuthal direction), or on the polar angle value to which all the motions correspond. The frequency shift occurs namely to the red, *i.e.* it can be treated as if the red Doppler shift would dominate over the gravitational (in this case, violet) one. Thus, using the red frequency shift of the standard signals coming from the mother craft, it is possible on the probe to determine the moment of crossing the horizon of any black hole. In this example, in contrast to the cosmological case, it is impossible to introduce a frame which is simultaneously co-moving with respect to both source and receiver of the signals: at the initial moment their time-like world lines coincide, so they cannot belong to the same congruence. This is however a meaningful example of application of a bilocal approach to reference frames, complementary with respect to the monad formalism.

3.4 The Dragging Phenomenon

The dragging phenomenon embraces a large variety of effects predicted for test particles moving in gravitational fields of rotating (and similar to them) sources, and it probably has a very profound and general physical nature. It is usually characterized as "dragging of local inertial frames", being considered — in the absolute majority of cases — in space-times admitting a rotating time-like Killing vector field ξ [see, *e.g.*, Misner, Thorne and Wheeler (1973)]. It is clear that in such space-times a privileged reference frame is easily introduced, namely that which coincides with the Killing congruence. Rotation of this reference frame describes then the dragging effects. We shall see that this rotation (essential stationarity of the space-time) is then an important prerequisite for existence of dragging, though usually in discussions of the dragging phenomenon nothing is mentioned concerning reference frames being used, maybe some traditional words about the local reference frames only (but without any elements of a sensible formalization of the reference frames theory). In this section, we attempt to give an introduction to consideration of the dragging phenomenon in the realm of this theory, as well as additional hints pointing at some prospects and needs in generalizing the very concept of dragging.

In a space-time admitting time-like Killing vector, there exists a preferred reference frame with

$$\tau = \frac{\xi}{\sqrt{\xi \cdot \xi}}. \qquad (3.4.1)$$

Such a Killingian reference frame can be interpreted as a co-moving one with the gravitational field (or with the physical system itself which produces this field), so that the frame is in the state of acceleration and rotation together with the space-time being considered (no deformation can obviously be present in this case). If this space-time is not static, but

---

[7] Already Brill (1972) pointed out that his and Schrödinger's approach is applicable (due to its invariance) even at points where the coordinates become degenerate.

stationary one, the dragging phenomenon occurs. It is then related to existence (in such a reference frame) of a gravitomagnetic field — the rotation vector ω [cf. Nordtvedt (1988), Jantzen, Carini and Bini (1992), as well as the gravitation-electromagnetism analogy following from the equation (3.2.14)]. If we change to another monad, $\tilde{\tau}$, this vector field may be of course transformed away, but since the preferred Killingian frame is supposed to be in a rotation, $\omega \neq 0$ has to be considered as possessing an objective meaning. We speak then about a rotating Killingian reference frame, whose rotation yields gravitomagnetic effects, and this is precisely the dragging phenomenon. Since there usually exists another (now, space-like) Killing vector field η which may be considered as complementary to ξ, a combination of these two fields (with constant coefficients) also represents a Killing vector field, the concrete choice of coefficients determining a hypersurface on which the corresponding rotation vector vanishes locally.[8]

It is worth giving here a concrete, but general enough example of how works the existence (or absence) of the rotation vector, from the point of view of a coordinated basis connected with τ. The Killing equation (1.2.24) can be rewritten as

$$\mathcal{L}_\xi g_{\mu\nu} \equiv g_{\mu\nu,\alpha}\xi^\alpha - g_{\alpha\nu}\xi^\alpha{}_{,\mu} - g_{\mu\alpha}\xi^\alpha{}_{,\nu} = 0. \tag{3.4.2}$$

One may choose coordinate (say, t) lines coinciding with the integral curves of the field ξ, $\xi \cdot \xi > 0$, together with a corresponding parametrization along these lines, so that in the new system of coordinates, $\xi = \partial_t \equiv \delta_t^\mu \partial_\mu$. Then (3.4.2) takes form $g_{\mu\nu,t} = 0$ (i.e., all metric coefficients are independent of the Killingian coordinate t). If there exist several independent Killing vector fields, the independence of all components $g_{\mu\nu}$ of all corresponding new (Killingian) coordinates may be achieved, only if these Killing vectors mutually commute. If not, the constancy of metric coefficients is realized in different systems of coordinates only (not unifiable into one system). We confine ourselves to a consideration of one single Killing vector field, ξ, and we will show that a non-rotating Killingian congruence corresponds to a static space-time, and *vice versa*. In the corresponding Killingian system of coordinates,

$$ds^2 = g_{tt}\, dt^2 + 2g_{ti}\, dt\, dx^i + g_{ij}\, dx^i dx^j,$$

$i, j = 1, 2, 3$ if $t = x^0$; all $g_{\mu\nu}$ are independent of t. Equivalently,

$$ds^2 = g_{tt}\left(dt + \frac{g_{ti}}{g_{tt}}dx^i\right)^2 + \left(g_{ij} - \frac{g_{ti}g_{tj}}{g_{tt}}\right)dx^i dx^j \tag{3.4.3}$$

or

$$ds^2 = (\tau_\mu dx^\mu)^2 + b_{\mu\nu}dx^\mu dx^\nu,$$

where[9]

$$\tau_\mu = \frac{g_{t\mu}}{\sqrt{g_{tt}}}, \quad b_{\mu\nu} = g_{\mu\nu} - \tau_\mu \tau_\nu.$$

---

[8]This analysis is most characteristic for studies of the ergosphere region, *e.g.*, of the Kerr space-time, and it corresponds to the local stationarity property of that space-time; cf. Hawking and Ellis (1973), p. 167. In the case of a pencil-of-light space-time, local stationarity is in fact a global property.

[9]The first of the expressions given below represents a monad τ in coordinates co-moving with the reference frame (Zel'manov's chronometric invariants formalism), while the second one simply duplicates (2.2.1).

A complete separation of t and x^i in (3.4.3) — orthogonalization of t-lines with respect to all other (here, spatial) coordinates lines — occurs when the expression $\zeta := \tau/\sqrt{g_{tt}} = dt + (g_{ti}/g_{tt})dx^i$ is a total differential. This case is known as the static field case (a simultaneous orthogonality of t with respect to other axes *and* t-independence of all $g_{\mu\nu}$). This corresponds to fulfilment of the condition

$$\left(\frac{g_{ti}}{g_{tt}}\right)_{,j} - \left(\frac{g_{tj}}{g_{tt}}\right)_{,i} = 0,$$

the same as vanishing of $A_{\mu\nu}$ (or, equivalently, of ω), i.e. $\xi \wedge d\xi = 0$. If $\omega \neq 0$, the space-time is *stationary* (but not static).

Thus we have come to the following chain of conclusions: If a Killing vector ξ is time-like, $\xi \cdot \xi > 0$, it is *always* possible to choose the coordinates in such a way that $\xi = \partial_t$; then $\xi \cdot \xi \equiv g_{tt}$, so that $\tau = \xi/\sqrt{g_{tt}}$. However, taking a 1-form $\zeta := \tau/\sqrt{g_{tt}} = \xi/g_{tt} = \xi/(\xi \cdot \xi)$,[10] one can introduce a new time coordinate $\tilde{t}$ such that $\zeta = d\tilde{t}$, *if and only if* $\omega = *(\xi \wedge d\xi) = 0$, and this is the static case.

3.4.1 Dragging in Circular Equatorial Orbits in the Kerr Space-Time

We shall consider first a very simple case of dragging, that which occurs in the equatorial plane of the Kerr space-time. The Kerr metric in the Boyer–Lindquist coordinates reads

$$ds^2 = \left(1 - \frac{2Mr}{\Sigma}\right)dt^2 - \frac{\Sigma}{r^2 - 2Mr + a^2}dr^2 - \Sigma d\vartheta^2$$

$$- \left(r^2 + a^2 + \frac{2Ma^2 r \sin^2\vartheta}{\Sigma}\right)\sin^2\vartheta d\phi^2 + 2\frac{2Mar\sin^2\vartheta}{\Sigma}d\phi dt, \qquad (3.4.4)$$

$\Sigma = r^2 + \cos^2\vartheta$, so that $\sqrt{-g} = \Sigma \sin\vartheta$; M is the central mass in units of length ($M = Gm$, G being the Newtonian gravitational constant) and a, the Kerr parameter (angular momentum *pro* unit mass, also of length's dimensionality). For a circular equatorial orbit ($r = \text{const.}$, $\vartheta = \pi/2$), radial component of the geodesic equation

$$0 = \frac{d}{ds}\left(g_{r\nu}\frac{dx^\nu}{ds}\right) = \frac{1}{2}g_{\alpha\beta,r}\frac{dx^\alpha}{ds}\frac{dx^\beta}{ds} \qquad (3.4.5)$$

gives enough information for solving the problem. Here $dx^\mu/ds = (dt/ds)\delta^\mu_t + (d\phi/ds)\delta^\mu_\phi$. There are two roots of the eq. (3.4.5),

$$\dot\phi_\pm = \left(a \pm \sqrt{\frac{r^3}{M}}\right)^{-1} \qquad (3.4.6)$$

---

[10] We use here in the same context both ξ and τ as vectors and covectors (and we do a kind of this throughout the book) with hope that the reader will not feel much inconvenience.

($\dot\phi := d\phi/dt$), so that

$$T_\pm := 2\pi \mid \dot\phi_\pm \mid^{-1} = 2\pi\left(\sqrt{\frac{r^3}{M}} \pm a\right) = T_N \pm \Delta T \qquad (3.4.7)$$

(we consider here a being a smaller term than $\sqrt{r^3/M}$ in a physically meaningful region of the motion). The main part of the period has here the standard Newtonian value T_N, while the dragging term is simply additive.[11] It is rather remarkable that it does not depend on the orbit's radius, mass of the attractive centre, and even on the gravitational constant, thus being suggestible from purely dimensional considerations. At the same time, it is worth stressing that this result is exact, and not approximate one [see Mitskievich and Pulido (1970), Mitskievich (1976), Mitskievich (1990); for an approximate approach in linearized gravity, see Mitskievich (1979)]. The fact that we have used here the coordinate (not proper) time and (coordinate) angle ϕ, does not make our results non-covariant, since these coordinates are the Killingian ones, thus invariantly reflecting geometric properties of the Kerr space-time. It is clear that the meeting point of two test particles moving in the opposite directions along the same orbit, then itself rotates with the angular velocity

$$\dot\phi_m = \frac{Ma}{Ma^2 - r^3}, \qquad (3.4.8)$$

and a simple calculation shows that (Mitskievich 1976) if for usual celestial bodies the effect is quite small, for super-dense objects (such as pulsars) it is very large: see the Table 3.1 in which the usual mass m (in grams) is used; we have taken the circular orbit's radius equal to $r = 3^{1/3}R$, where R is radius of the real attracting centre, L angular momentum of the central body, D the meeting point drift in seconds of arc per century (the Schwarzschild precession Δ in the same units being also given as illustrative data, R meaning then the large semi-axis of an elliptic orbit).

Another analogous dragging effect is that of a Kerr gravitational lens when two light rays passing by the Kerr centre in its equatorial plane with the same impact parameter but on opposite sides of the centre, are deviated differently by the gravitational field of the rotating source, and they are focused off the "straight" line going through the centre parallel to the initial common direction of the rays (a transverse shift of the focus of these rays). The magnitude of this shift from the above "straight" line is *equal* to the Kerr parameter (if measured in the units of distance), this being however an approximate result which becomes more precise for large values of the impact parameter (Mitskievich and Gupta 1980, Mitskievich and Uldin 1983, Mitskievich and Cindra 1988).

Let us now apply definition of the rotation covector ω (2.3.5) to the Killingian frame in the Kerr space-time. In fact, this is the only object which could be connected with the quantitative manifestations of the dragging phenomenon. Since in a stationary space-time

---

[11] In the Kerr–Newman space-time (of a rotating electrically charged black hole, Q being its charge in units of length) a similar exact solution takes place, the only difference from this one being $T_N = 2\pi r^2/\sqrt{(Mr - Q^2}$, with the same further remarks applicable to it (Mitskievich and López-Benítez 2004).

Table 3.1: Dragging in the Kerr space-time

Central body	m, g	R, cm	L, $\frac{g \cdot cm^2}{s}$	T_N, s	D	Δ
Sun	$2 \cdot 10^{33}$	$7 \cdot 10^{10}$	$1.3 \cdot 10^{49}$	$2 \cdot 10^4$	600	10^6
Earth	$6 \cdot 10^{27}$	$6 \cdot 10^8$	$7 \cdot 10^{40}$	$9 \cdot 10^3$	4.4	670
Jupiter	$2 \cdot 10^{30}$	$7 \cdot 10^9$	$7 \cdot 10^{45}$	$2 \cdot 10^4$	280	10^4
Rapidly rotating class B star	$3 \cdot 10^{34}$	$3.5 \cdot 10^{11}$	$2 \cdot 10^{53}$	$5 \cdot 10^4$	$7.6 \cdot 10^4$	10^6
Pulsar at a breaking point	10^{33}	10^6	$4 \cdot 10^{48}$	10^{-3}	93 $\frac{rad}{sec}$	750 $\frac{rad}{sec}$

(as also in a static one) $\xi = \partial_t$, (3.4.1) reads

$$\tau = \frac{g_{\mu t}}{\sqrt{g_{tt}}} dx^\mu;$$

thus (2.3.5) yields

$$\omega = \frac{g_{\kappa t}}{2 g_{tt}} g_{\mu t, \lambda} E^{\kappa \lambda \mu}{}_\nu dx^\nu.$$

First, we observe that κ, $\mu = t$, ϕ and $\lambda = r$, ϑ. Hence, $\nu = \vartheta$, r (complementary to λ). Then it is clear that on the equator, ω is directed along ϑ, i.e. the rotation occurs in the $(\pm) \phi$ direction. This is quite natural and corresponds to a similar rotation of the source of Kerr's field. The general exact expression is actually

$$\omega = \frac{1}{2} E^{tr\vartheta\phi} \left[\left(g_{t\phi,\vartheta} - \frac{g_{t\phi}}{g_{tt}} g_{tt,\vartheta} \right) \partial_r - \left(g_{t\phi,r} - \frac{g_{t\phi}}{g_{tt}} g_{tt,r} \right) \partial_\vartheta \right] \quad (3.4.9)$$

(contravariant vector representation).

On the equator ($\vartheta = \pi/2$), the rotation vector (3.4.9) takes the form

$$\omega_{\mathrm{eq}} = \frac{\gamma m a / r^4}{1 - 2\gamma m/r} \partial_\vartheta, \quad \sqrt{\omega_{\mathrm{eq}}^2} = \frac{\gamma m a}{r^2(r - 2\gamma m)}. \quad (3.4.10)$$

We have to compare this absolute value (which is a *scalar*) with (3.4.8) (which is neither a scalar nor component of a vector). They coincide (up to the sign) when the field is weak (i.e. $r \gg \gamma m$, $r \gg a$). Why this coincidence realizes for a weak field only? The answer could be that (1) we compare quantities with different geometric natures (see our remarks in the parentheses above) which may become similar in the vicinity of flat space-time only; (2) the very description of the dragging effect as a mean value of two rotation frequencies is highly tentative, and could serve well in the weak field approximation, but not in the general case.

3.4.2 An Orbit Shift in the Taub–NUT Space-Time

A closely analogous, but much more exotic dragging effect is present in the Taub–NUT space-time — the field of a gravitational dyon (Mitskievich 1981, Mitskievich 1983). Here a gravitoelectric field of the Schwarzschild type is accompanied by a gravitomagnetic field, similar to the magnetic field of a magnetic monopole. If one considers in this field circular orbits of test masses, one finds that the field centre cannot be in the plane of such an orbit, but it is shifted in the direction perpendicular to the plane, by a distance proportional to the value of the orbital angular momentum and to the NUT parameter l, the direction of the shift being the same as direction of the angular momentum L for $l > 0$, and opposite to this direction for $l < 0$.

Let us consider this case in some detail. While the gravitational mass may be called gravitoelectric charge, the NUT parameter l is similar (to certain extent) to gravitomagnetic monopole charge (from the structure of Weyl's tensor the differences are fairly obvious). The vacuum Taub–NUT metric is $ds^2 = \frac{\Delta}{\Sigma}(dt + 2l\cos\vartheta d\phi)^2 - \frac{\Sigma}{\Delta}dr^2 - \Sigma\left(d\vartheta^2 + \right.$ $\left.+\sin^2\vartheta d\phi^2\right)$, where $\Delta(r) = r^2 - 2Mr - l^2$, $\Sigma(r) = r^2 + l^2$; see for more details (Carmeli 2000, Stephani *et al.* 2003).

We now consider another analogue of electromagnetic Zeeman effect (different from motion of a test charge in a combination of the Coulomb and magnetic dipole fields which was the prototype of dragging in the Kerr and Kerr–Newman space-time studied in the preceding subsection), similar to the motion of an electrically charged point-like mass around a centre possessing both electric and magnetic monopole charges. Thus we consider a circular motion of a (neutral) test mass about the Taub–NUT centre; like in the electromagnetic case, the orbit has to be centred on the z axis and not precisely on the origin. Then we have to use the conditions $dr = 0 = d\vartheta$, thus r- and ϑ-components of the geodesic equation, $\frac{d}{ds}\left(g_{\mu\nu}\frac{dx^\nu}{ds}\right) = \frac{1}{2}g_{\alpha\beta,\mu}\frac{dx^\alpha}{ds}\frac{dx^\beta}{ds}$, yield

$$\tan\vartheta = \pm\frac{1}{2l}\sqrt{\frac{g_{tt,r}}{2r}\left(\frac{\Sigma^2}{\Delta} - 8l^2\right)}, \qquad (3.4.11)$$

where $g_{tt,r} = 2\frac{Mr^2 + 2l^2 r - Ml^2}{\Sigma^2}$. When $l = 0$, the orbit is, of course, centred on the origin ($\tan\vartheta = \infty$), but in the Taub–NUT case proper, it lies above or under the origin depending on the relative sign of l and the test particle's angular momentum, as one can see from the relation (3.4.11) plus an elementary consideration of two conservation laws (those of energy $\mathcal{E}$ and angular momentum $\mathcal{L}$, both taken per unit rest mass). Another form of (3.4.11) then reads $\cos\vartheta = -\frac{2l\mathcal{E}}{\mathcal{L}}$ (Mitskievich 1983, 1996).

3.4.3 Dragging in the Space-Time of a Pencil of Light

Now let us consider a stationary case of the pencil-of-light space-time (Mitskievich 1981a):

$$ds^2 = dt^2 - d\rho^2 - dz^2 - \rho^2 d\phi^2 + 8\gamma\epsilon\ln(\sigma\rho)(dt - dz)^2. \qquad (3.4.12)$$

Here ϵ is the energy (or, equivalently, momentum z-component) linear density, while σ serves simply for making the argument of logarithm dimensionless (in a non-stationary case, both ϵ and σ are treated as arbitrary functions of the retarded time $t - z$); γ is the Newtonian gravitational constant.

The already mentioned stationarity property of the metric (3.4.12) is non-trivial. There always exist Killing vectors $\xi = \partial/\partial t$ and $\eta = \partial/\partial z$ (as well as $\partial/\partial\phi$, of course). The squares of ξ and η are not sign-definite, so that t is not always a time-like coordinate, and z not always a space-like one. In the region $-1 < 8\gamma\epsilon \ln \sigma\rho < +1$ the coordinates are respectively time- and space-like. When $8\gamma\epsilon \ln \sigma\rho < -1$ both coordinates are space-like, and when $8\gamma\epsilon \ln \sigma\rho > +1$, they are time-like. But at any point one may take linear combinations (with constant coefficients) of these two Killing vectors, yielding two new Killing vectors being correspondingly time-like and space-like. Moreover, the band $-1 < 8\gamma\epsilon \ln \sigma\rho < +1$ may be then shifted to every position (retaining its finite width) if the following coordinate transformation is used:

$$t = (1 - L)t' + Lz', \quad z = -Lt' + (1 + L)z', \quad L = -4\gamma\epsilon \ln(\sigma'/\sigma),$$

while

$$8\gamma\epsilon \ln(\sigma\rho) \rightarrow 8\gamma\epsilon \ln(\sigma'\rho).$$

From the geodesic equation it is easy to conclude about existence in the pencil-of-light field of the phenomenon of dragging; at the same time, the time-like Killing vector congruence (in the above-mentioned band) rotates, so that in the privileged frame there exists a gravitomagnetic field (alongside with a gravitoelectric one). With λ as a canonical parameter, the first integrals of motion are

$$(1 + 8\gamma\epsilon \ln(\sigma\rho))\frac{dt}{d\lambda} = \alpha, \quad (1 - 8\gamma\epsilon \ln(\sigma\rho))\frac{dz}{d\lambda} = \beta,$$

$$\rho^2 \frac{d\phi}{d\lambda} = \mu,$$

α, β and μ being the integration constants. For ρ we have the equation

$$\frac{d^2\rho}{d\lambda^2} = -\frac{4\gamma\epsilon}{\rho}(\alpha - \beta)^2 + \frac{\mu^2}{\rho^2}$$

[it can be easily integrated to a first integral, but this reduces to the squared interval (3.4.12)]. In the right-hand side, the first term describes gravitoelectric attraction to z axis, and the second one, the centrifugal repulsion. The dragging effect is in this case compensated by the (conserved) z-component of linear momentum of the test particle.

If the test particle even did momentarily move without changing its z-coordinate, being however not on a circular orbit in a plane perpendicular to z axis, this particle would be dragged in the next moment along this axis, the sign of the dragging being determined by the expression

$$(\alpha - \beta)8\gamma\epsilon\frac{\Delta\rho}{\rho_0}$$

where ρ_0 is the value of ρ when, momentarily, $dz/d\lambda = 0$, and $\Delta\rho$ is the deviation of ρ from ρ_0.

It is worth mentioning that a photon (light-like geodesic) moving parallel to z axis in its positive direction, does not interact with the pencil-of-light field at all [see for a general discussion Mitskievich (1989)], but a photon moving in the opposite sense will fall subsequently onto the pencil of light (z axis). This fact of no interaction holds not only for test objects in the pencil-of-light field, but it is also true for self-consistent problems with light-like moving objects in general relativity. Thus we come to exact additivity[12] of gravitational fields of parallel (not anti-parallel) pencils of light; this being an exact generalization of the known approximate result obtained by Tolman, Ehrenfest and Podolsky (1931). This effect is however even quite general since it acts not only in classical, but also in quantum theory. Moreover, it has to do with all types of objects (particles and fields) performing globally light-like motion in the same direction under any physical interactions, and it occurs also in the form of asymptotic situation as vanishing of influence of these interactions when the kinetic energy of both counterparts (in fact, the energy of as many objects as one would wish) becomes individually very large in comparison to the respective rest energy of these counterparts.

Another case of the pencil-of-light field involves not only luminal (null) motion of the field's source, but its rotation too, this rotation being considered as an analogue of polarization of an infinitesimally thin beam of light. The corresponding metric reads

$$ds^2 = 2d(t-z)[d(t+z) + (4\gamma\epsilon \ln(\sigma\rho) - a\phi)\, d(t-z)] - d\rho^2 - \rho^2 d\phi^2 \qquad (3.4.13)$$

which may be modified to

$$ds^2 = 2d(t-z)[d(t+z) + 4\gamma\epsilon \ln(\sigma\rho)d(t-z) + (t-z)ad\phi] - d\rho^2 - \rho^2 d\phi^2, \qquad (3.4.14)$$

where a is a new constant corresponding to the "polarization" (in a more general case considered from another point of view in chapter 5, instead of the product $(t-z)a$ there stands an arbitrary function $g(t-z)$; then the space-time becomes however non-stationary). The metric (3.4.13) describes a stationary gravitational field, thus permitting application of the standard approach to the dragging phenomenon. Here, as well as in the alternative version (3.4.14), we see that the rotational motion when coexisting with the luminal one, inevitably leads to a *dragging of dragging* effect, that is, to appearance of the product term $dzd\phi$, in addition to $dtd\phi$, $dtdz$ being already present by the virtue of the pencil-of-light nature of the initial metric. This is not a merely accidental presence of all three terms in a concrete solution, but *inevitability* of the presence of them all, if any two of them are already present in a solution.

The simplest approach to consideration of dragging in the pencil-of-light space-time consists of a calculation of the (proper) time derivatives of coordinates of a free test object. We mean here not covariant, but simply partial derivatives which manifest the tendency of varying the coordinates (or their derivatives, if only the corresponding higher order derivatives become non-vanishing). Thus, considering as an initial condition an instantaneous

---

[12]This means, of course, in a broader approach, 'additivity' in a certain general sense, which manifests itself as the 'trivial' additivity when the Kerr–Schild form of metric is applicable (for example, in our consideration of the pencil-of-light case).

state of "rest", we see that the geodesic equation yields (through the second order derivatives) a (non-covariant) acceleration towards the pencil-of-light source (in the $-\rho$ direction); these (and possibly the third order) derivatives show, moreover, that *a tendency* of dragging exists, acting in the ϕ (polarization) and z (orientation of the light-like motion of the pencil of light) directions. We encounter then, as manifestations of dragging in this special kind of field, the *germs* of acceleration [see Mitskievich and Kumaradtya (1989)].

Let us discuss this problem quantitatively using a more general form of the *spinning pencil of light* (SPL) metric,

$$ds^2 = 2dv\left(du + k(v)\ln\sigma\rho dv + g(v)d\phi\right) - d\rho^2 - \rho^2 d\phi^2$$

(a possible dependence $\sigma(v)$ is inessential); here the variables v and u correspond to $t - z$ and $t + z$ in (3.4.13).

Since the metric coefficients are independent of u and ϕ, one immediately arrives at two first integrals of motion,

$$g_{vu}\frac{dv}{d\lambda} = \alpha > 0, \quad g_{v\phi}\frac{dv}{d\lambda} + g_{\phi\phi}\frac{d\phi}{d\lambda} = -\beta.$$

The further exact integration of the geodesic equation is in general impossible, but we are not interested in obtaining approximate solutions (*e.g.*, using perturbative procedures), so another way to deal with the dragging phenomenon should be chosen. We shall consider only *germs* — the tendency of test particles motion, *i.e.*, the first non-vanishing higher order derivatives of the particles' spatial coordinates when the initial state of motion is given. This initial motion is most naturally chosen as a state of momentary rest (if the test particle has non-zero rest mass, thus $d\lambda = ds$). Then

$$\left(\frac{d\rho}{ds}\right)_0 = \left(\frac{d\phi}{ds}\right)_0 = \left(\frac{dz}{ds}\right)_0 = 0.$$

We may now express z in analogy with the Peres wave: $z = (1/2)(u - v)$ [*cf.* also (3.5.10)]. Then

$$\left(\frac{dv}{ds}\right)_0 = \left(\frac{du}{ds}\right)_0.$$

The remaining components of the geodesic equation yield

$$\left(\frac{d^2\rho}{ds^2}\right)_0 = -\frac{k\alpha^2}{\rho}, \tag{3.4.15}$$

$$\left(\frac{d^2 z}{ds^2}\right)_0 = -\frac{g\dot{g}\alpha^2}{2\rho^2} - \frac{\alpha^2 k\ln\sigma\rho}{2}, \tag{3.4.16}$$

$$\left(\frac{d^2\phi}{ds^2}\right)_0 = -\frac{\dot{g}\alpha^2}{\rho^2}, \tag{3.4.17}$$

where the right-hand side quantities are also taken at the initial moment (in the sense of $(...)_0$). Since $k = 8\gamma\epsilon_0 > 0$ (the linear energy density of the SPL source must be positive), (3.4.15) describes attraction of the test particle to the SPL. The noncovariant accelerations in the directions of z and ϕ, (3.4.16) and (3.4.17), may be either positive or negative depending on values and signs of g, $\dot{g}$, and k, as well as on ρ. The sign of dragging in the ϕ direction coincides with that of $\dot{g}$, and this means that not the function g, but its first derivative is directly related to the angular momentum of the SPL which is responsible for dragging in this direction. Moreover, if g is constant, it can be transformed away by merely introducing $\tilde{u} = u + g\phi$, so that only $\dot{g}$ can have a physical meaning.

As to the motion of a lightlike particle, its initial state should be now chosen in a different way, since the former initial conditions are incompatible with $ds^2 = 0$. So we shall take only two of these conditions,

$$\left(\frac{d\rho}{d\lambda}\right)_0 = \left(\frac{d\phi}{d\lambda}\right)_0 = 0.$$

Inserting them into ds^2, we have

$$\left(\frac{dv}{d\lambda}\right)_0 \left(\frac{du}{d\lambda} + k\ln\sigma\rho\frac{dv}{d\lambda}\right)_0 = 0,$$

which leads to two possibilities:

$$\left(\frac{dv}{d\lambda}\right)_0 = 0$$

and

$$\left(\frac{du}{d\lambda} + k\ln\sigma\rho\frac{dv}{d\lambda}\right)_0 = 0. \tag{3.4.18}$$

The first one corresponds to absence of interaction between parallelly moving lightlike objects (in fact, also not test ones). The alternative case (3.4.18) does not admit $(dv/d\lambda)_0 = 0$ since such a case would correspond to a world point (an event) and not a world line. Combining the condition (3.4.18) with the geodesic equation (including the first integrals of motion), we come to the relations

$$\left(\frac{d^2\rho}{d\lambda^2}\right)_0 = -\frac{k\alpha^2}{\rho},$$

$$\left(\frac{d^2\phi}{d\lambda^2}\right)_0 = \frac{\dot{g}\alpha^2}{\rho^2},$$

$$\left(\frac{d^2u}{d\lambda^2}\right)_0 = -\frac{g\dot{g}\alpha^2}{\rho^2} - k\alpha^2\ln\sigma\rho.$$

These accelerations are in fact the same as in the case of a massive test particle, though we have here u instead of z. We see that a photon moving parallel to a SPL does not interact with it (in particular, it does not feel dragging in the ϕ direction), while a photon moving antiparallel to it both falls onto the SPL and starts to rotate in the ϕ direction.

3.4.4 Other Dragging Effects

In this comparatively short account on dragging we can but mention many other aspects of dragging, though they can be easily related to the reference frames' theory. These include, first of all, effects of change of orientation of test bodies possessing multipole (beginning with the dipole) moments (thus without spherical symmetry, or scalar property). This is, in particular, the Schiff effect (precession of a gyroscope). Effects of this type are sometimes classified as non-dragging ones, since they are connected with a local rotation of test particles without necessarily a translational motion, but we cannot consider this distinction as essential [see, *e.g.*, Sakina and Chiba (1980), Marck (1983), Tsoubelis, Economou and Stoghianidis (1987), Tsoubelis and Economou (1988)]. This motion includes precession-like manifestations not only of dragging, but also of the tidal forces *etc* [see also Ashby and Shahid-Saless (1990)]. The mentioned papers contain additional bibliography on the subject.

A specific character of these effects includes arbitrariness of the gyroscope world line which can be assigned artificially (usually, one takes a geodesic which is a good approximation for a freely moving gyroscope). As another example one may consider a gyroscope suspended at some fixed point over the surface of the Earth.

To this type of dragging belong effects involving non-scalar (in particular, rotating) test bodies which do not change their orientation (relative to the directions determined by symmetries of the gravitational field) in course of the motion. Thus no precession-type effects are now considered. For example, a spinning particle can hover over a Kerr source pole if

$$M(z^2 - a^2) = -2aS \frac{3z^2 - a^2}{z^2 + a^2} \left(1 + \frac{2m|z|}{z^2 + a^2}\right).$$

Here z is position of the test particle over the pole and S, spin of the particle. This effect represents the spin-spin interaction (Epikhin, Pulido and Mitskievich 1972). An example of the spin-orbital interaction can be given for a circular motion in the Schwarzschild field of a pair of spinning test particles with antiparallel spins orthogonal to the plane of their common orbit (see the same publication). Their periods of revolution are approximately given by the formula

$$T_\pm = 2\pi \left(\sqrt{\frac{r^3}{\gamma m}} \mp \frac{3S}{2mc^2}\right);$$

hence, these two particles will chase one another with different linear velocities: the difference,

$$\Delta v = \frac{3\gamma m S}{Mc^2 r^2},$$

leads to a drift

$$\Delta l = 6\pi \sqrt{\frac{\gamma m}{rc^2}} \frac{S}{Mc}$$

of one particle with respect to the other per one revolution. Then in order to attain a drift of some 1 Ångström per one revolution in a low orbit about the earth, the gyroscopes should spin with an angular velocity

$$\Omega \approx 10^5 R^{-2} \, sec^{-1},$$

where R (cm) is the characteristic radius of inertia of the gyroscope. Some similar dragging (or quasi-dragging) effects, but in a quantum-mechanical description, were discussed in the monograph (Mitskievich 1969). An elegant study of problems related to reference frames, gravitation and dragging in quantum physics, see also in (Schmutzer 1975, 2004).

3.5 More General Gravitoelectromagnetic and Gravitoelectric Phenomena

In this section we consider some manifestations of gravitoelectromagnetism whose nature is different of (or more general than) that of a pure dragging. First of all, let us focus attention on a close analogue of the dragging phenomenon (not merely in mechanics but essentially in the electromagnetic field theory) connected with results of DeWitt (1966) and supporting the conclusion about inextricable links between gravitomagnetic and gravitoelectric effects in general relativity. These results consist of two parts:

(1) When a conductor is placed into a superposition of electric and gravitoelectric (Schwarzschild type) fields, or is considered (in the presence of an electric field) in a translationally accelerated reference frame, inside such a conductor there vanishes not the electric field, but a certain combination of that and of the gravitoelectric field,

$$eE - mG = 0, \qquad (3.5.1)$$

e and m being the charge and mass of electron (inside the conductor, a "free" electron gas has to be considered). Hence, in the conductor there must exist an electric field exactly corresponding to the acceleration of the monad congruence which describes the co-moving frame of this conductor.

(2) When a superconductor is placed into a superposition of a magnetic and gravitomagnetic fields (one may consider, *e.g.*, the rotating Earth's field — say, the Lense–Thirring or Kerr one), or the superconductor belongs to a rotating reference frame, inside this superconductor there vanishes not the magnetic field, but its combination with the gravitomagnetic field,

$$eB + m\omega = 0. \qquad (3.5.2)$$

So it is usually said that in a rotating superconductor a corresponding magnetic field is generated.

We propose to write in the case of a superconductor a general covariant equation

$$d(mu + eA) = 0 \qquad (3.5.3)$$

[this is a general covariant form of eqs. (3.5.1) and (3.5.2); *cf.* also (3.1.3)]. One may characterize this effect as *dragging of electric and magnetic fields by the corresponding gravitational — or inertial — fields*, *cf.* below (in section 4.3) kinematic "source" terms which appear in Maxwell's equations in non-inertial frames, and which are analogous to the Coriolis and centrifugal inertial forces terms in the equations of classical mechanics.

It is clear that this new type of dragging makes it easier to measure gravitational fields, since it reduces gravitational measuring problems to the corresponding electromagnetic ones. Some proposals along these lines were made, *e.g.*, by Papini (1966, 1969), although no general covariant exact formulation was ever proposed [see also a consideration of the superconductivity theory in general relativity by Meier and Salié (1979) and Salié (1986)].

It is also obvious that our approach makes it possible as well to consider non-stationary gravitational fields (and non-uniformly accelerated and rotating frames). Thus we predict that when a gravitational wave overrides a superconductor sample, corresponding electromagnetic oscillations are to be observable inside it, the measuring problem being reduced to that of detecting electromagnetic oscillations inside a superconductor. This is the case when no time-like Killing congruence could be present, which in the general opinion is however an indispensable premise for consideration of the dragging phenomenon itself. So, losing the symmetry in the sense of time, do we really lose the very thread of our argument?

It was however so easy to lose this tiny thread which does exist in highly idealized cases only. Adding merely a "grain" of impurity to the pure symmetry (here: stationarity), we extinguish the latter completely. But if a *physical effect* — or *phenomenon* — is so very vulnerable, it either does not exist at all, or its formulation is to be crucially revised. For example, in the Kerr field we encounter a great multitude of dragging effects; so, if from some distant source an infinitesimally weak gravitational wave comes to this Kerr field region, could this lead to a total and instantaneous breakdown of all such effects? Is it not much more likely that our *definition* of the dragging phenomenon (however natural it might seem to be) is in fact doomed to a radical revision? A symmetry is usually a good tool which makes it possible to dig up some principal results in the quickest and most elegant way, but we have not to reject a possibility of generalizing these results beyond the limits of that symmetry approach, especially if the latter is so critically unstable. Description of the dragging phenomenon becomes very vague without the use of ideas of symmetry, because we have then no preferred reference frame — a very typical situation in general relativity. It is however quite plausible that the effects considered above in a stationary case, should hold also for fields essentially dependent on time. Thus in a cavity inside superconducting medium, there should be generated electromagnetic oscillations when a gravitational radiation pulse would override that region, so that the oscillations should imitate that pulse. Such a feature (one may speak about a tendency) is characteristic for gravitation itself (Mitskievich 1983). Somebody could insist that the notion of dragging is foreign to such effects, and probably it is not so important what name to give to the phenomenon the boundary of which is so diffuse. We are inclined nevertheless to speak about generalized manifestations of dragging, — let it be, for example, a combination of gravitoelectric and gravitomagnetic fields and corresponding effects. In fact, to *every* electromagnetic effect should correspond a certain gravitational effect (though not *vice versa*) which should be named a gravitoelectromagnetic one, a class of such effects clearly pertaining to the dragging phenomenon area.

Secondly, let us study a simpler purely gravitoelectric case. The Reissner–Nordström metric

$$ds^2 = \left(1 - \frac{2M}{r} + \frac{Q^2}{r^2}\right) dt^2 - \frac{dr^2}{1 - \frac{2M}{r} + \frac{Q^2}{r^2}} - r^2\left(d\vartheta^2 + \sin^2\vartheta d\phi^2\right) \quad (3.5.4)$$

describes a space-time which, of course, does not reveal any gravitomagnetic properties in the invariant sense (the time-like Killing vector is non-rotating), thus its gravitational field is essentially gravitoelectric. In the region where the metric tensor differs only slightly from the Minkowski one, the Newtonian gravitational potential $\Delta\Phi_N$ can be introduced, and its electromagnetic source is two times greater than it should be in the non-relativistic case [see Mitskievich (2003)]. It is interesting that this fact leaves its mark on the general relativistic (gravitoelectric) properties of the space-time.

In order to study these properties, consider here the behavior of an electrically neutral test particle in the Reissner–Nordström space-time. If such a particle performs a purely radial motion ($d\phi = 0 = d\vartheta$), the geodesic equation yields the first integrals $\frac{dt}{ds} = \mathcal{E}/g_{tt}$ and $\frac{dr}{ds} = \pm\sqrt{\mathcal{E}^2 - g_{tt}}$ where the integration constant $\mathcal{E}$ is the particle's energy per unit of its rest mass, the sign $\pm$ corresponding to the motion from the centre or toward it. One may then infer that there should be a spherical equilibrium surface around the position of central mass at the radius $r_0 = Q^2/M$ (an elementary extremum problem), such that on it an initially resting test particle would continue to be at rest; moreover, a small radial deviation from this position would result in an oscillatory motion around it. But where this equilibrium surface is positioned with respect to the Reissner–Nordström horizons (the event and Cauchy ones)? These horizons correspond to the radii $r_{H_\pm} = M \pm \sqrt{M^2 - Q^2} = M\left(1 \pm \sqrt{1 - Q^2/M^2}\right)$, thus $r_{H_\pm} = M\left(1 \pm \sqrt{1 - r_0/M}\right)$. This means that $r_0 = r_{H_\pm}(2 - r_{H_\pm}/M) \equiv r_{H_+}r_{H_-}/M$. In other words, since $r_{H_-} < M < r_{H_+}$, it follows that the equilibrium surface is located *between* the horizons: $r_{H_-} < r_0 < r_{H_+}$, where the r coordinate is time-like and not space-like at all. (We consider here the most complicated and realistic case of $Q < M$, thus $r_0 < M$, i.e. when the horizons r_{H_+} and r_{H_-} exist and do not coincide one with another.)

And nevertheless, the equilibrium surface has a proper physical meaning. First of all, the Komar (1958, 1959) conservation law

$$\left(\sqrt{-g}\, T^\mu_\nu \xi^\nu\right)_{,\mu} \equiv \sqrt{-g}\, \left(T^\mu_\nu \xi^\nu\right)_{;\mu} = 0, \quad (3.5.5)$$

(where $\sqrt{-g}$ is the metric tensor determinant and ξ^μ, a Killing vector of this space-time) gives an evaluation of the integral conserved quantity U_{em} corresponding to both ξ^μ and T^μ_ν (the stress-energy tensor, here of the Reissner–Nordström electromagnetic field). When we consider ξ^μ as the time-like Killing vector (equal to ∂_t), the energy U_{em} contained in the concentric volume corresponding to $r_a \leq r < \infty$, is $U_{em} = 4\pi \int_{r_a}^{\infty} T^t_t r^2 dr$. With the electromagnetic four-potential $A = \frac{Q^2}{r} dt$ and the field tensor $F = dA$, $T^\mu_\nu = -\frac{1}{4\pi}\left(F_{\nu\lambda}F^{\mu\lambda} - \frac{1}{4}F_{\alpha\beta}F^{\alpha\beta}\delta^\mu_\nu\right)$ yields $U_{em} = \frac{Q^2}{2r_a}$. Since we take $c = 1$ and consider as the contribution of electromagnetic field to gravitational mass the quantity $2U_{\text{em}}$ [see Mitskievich (2003)], the (gravitational) masses of the Reissner–Nordström physical system observed from infinity, M, and from the radius r, $M(r)$, are mutually related as

$$M = M(r) + \frac{Q^2}{r} = \text{const.} \tag{3.5.6}$$

We immediately see that $M(r_0) = 0$, $M(0) = -\infty$, and $M(\infty) = M$. These conclusions exactly conform to the notion of the equilibrium surface and the known fact that the Reissner–Nordström singularity acts with an infinite repulsion on neutral test particles as well as on photons (the latter ones suffer infinite Doppler redshift when they arrive at it). At the same time the first of these conclusions directly confirms in the exact general relativity the result found in (Mitskievich 2003) for the Newtonian-type approximation.

The last step to be made in our treatment of the Reissner–Nordström space-time, is to solve the problem of the time-like behavior of r coordinate between the horizons *versus* the concept of equilibrium surface. To this end we introduce synchronous coordinates in this space-time. Remember that synchronous coordinates (here, $\{T, R, \vartheta, \varphi\}$) are those whose T-lines are non-rotating time-like geodesics parameterized by the proper time, and the spatial coordinates lines are lying in the global subspace orthogonal to T-lines; moreover, in our case all three families of these latter lines also are mutually orthogonal. Our considerations are based on free (geodesic) motion of neutral test particles with its first integrals given above. Then

$$dT = \mathcal{E} dt + \frac{\sqrt{\mathcal{E}^2 - g_{tt}}}{g_{tt}} dr, \quad dR = \frac{dt}{\mathcal{E}} + \frac{dr}{g_{tt}\sqrt{\mathcal{E}^2 - g_{tt}}}, \tag{3.5.7}$$

while the angular coordinates remain unchanged. Thus

$$\mathcal{E}^2 dR - dT = dr/\sqrt{\mathcal{E}^2 - g_{tt}}, \tag{3.5.8}$$

so that r is a function of the combination $\mathcal{E}^2 R - T$, while

$$\begin{pmatrix} \partial T/\partial t & \partial T/\partial r \\ \partial R/\partial t & \partial R/\partial r \end{pmatrix} = \begin{pmatrix} \mathcal{E} & \mathcal{F}/g_{tt} \\ \mathcal{E}^{-1} & (\mathcal{F} g_{tt})^{-1} \end{pmatrix},$$

$$\begin{pmatrix} \partial t/\partial T & \partial t/\partial R \\ \partial r/\partial T & \partial r/\partial R \end{pmatrix} = \begin{pmatrix} \mathcal{E}/g_{tt} & -\mathcal{E}\mathcal{F}^2 \\ -\mathcal{F} & \mathcal{E}^2 \mathcal{F} \end{pmatrix}.$$

Here $\mathcal{F}(r) = \sqrt{\mathcal{E}^2 - g_{tt}}$. In this simplest case of synchronous coordinates the Reissner–Nordström metric reads

$$ds^2 = dT^2 - \mathcal{E}^2 \left(\mathcal{E}^2 - g_{tt}\right) dR^2 - d\Omega^2. \tag{3.5.9}$$

Let us now consider the transformation law of T_T^T (temporal-temporal component of the electromagnetic stress-energy tensor in synchronous coordinates): $T_T^T = \frac{1}{g_{tt}} \left[\mathcal{E}\left(T_t^t - T_r^r\right)\right] +$

T_r^r. Since $T_\beta^\alpha = \frac{Q^2}{8\pi r^4}\left(\delta_t^\alpha \delta_\beta^t + \delta_r^\alpha \delta_\beta^r - \delta_\vartheta^\alpha \delta_\beta^\vartheta - \delta_\phi^\alpha \delta_\beta^\phi\right)$, $T_T^T = T_t^t = \frac{Q^2}{8\pi r^4}$, so that the electromagnetic energy integrals from r_0 to ∞ along the R and r lines respectively, coincide in both systems: $\int T_T^T \mathcal{E} \sqrt{\mathcal{E}^2 - g_{tt}} r^2 \sin\vartheta dR d\vartheta d\phi = \int T_t^t r^2 \sin\vartheta dr d\vartheta d\phi$. This fact settles the problem most obviously when the integration constant $\mathcal{E}$ is equal to one (the test particles whose motion models the synchronous coordinates, start from the rest at infinity).

It is plausible that this example of gravitoelectric effect can be also interpreted as a hint to a special rôle of gravitation in effective renormalization of mass (energy) in a wide area of the (at least, classical) field theory.

Chapter 4

The Maxwell Field Equations

4.1 The Four-Dimensional Maxwell Equations

To the end of the future applications of Maxwell's equations, we give here different representations of these equations in the four-dimensional form (without considering specific reference frames), with some relevant comments.

We recall first the structure of the Lagrangian density [see Landau and Lifshitz (1971, 1973); Synge (1965); Mitskievich (1969)]

$$\mathfrak{L}_{\text{em}} = -\frac{\sqrt{-g}}{16\pi} F_{\mu\nu} F^{\mu\nu} \tag{4.1.1}$$

which yields Maxwell's equations through the standard variational principle. This Lagrangian density can be reshaped by singling out a divergence term:

$$F_{\mu\nu} F^{\mu\nu} = (2 A_\nu F^{\mu\nu})_{;\mu} - 2 A_\nu F^{\mu\nu}{}_{;\mu} = *d(A \wedge *F) - A \cdot \delta F. \tag{4.1.2}$$

The divergence (as usually) does not contribute to the Lagrange–Euler equations, but the variational principle should now be regarded as applied to a Lagrange density which includes higher (second) order derivatives. The marvel of this reshaping is that the corresponding action integral can be varied according to the Palatini method known in the gravitation theory (Palatini 1919), *i.e.* when both electromagnetic four-potential and field tensor are varied independently. Then the variation with respect to the four-potential yields the Maxwell equations, and that with respect to the field tensor, the standard expression for this tensor as a four-dimensional curl of the four-potential (*cf.* the situation in the gravitational field theory). As it is in the gravitational case, one has herewith to assume that Lagrangian densities of the other fields do not contain the electromagnetic field tensor, so that if it entered those Lagrangians before the reshaping (4.1.2) (which would be by itself exotic enough), then after passing to (4.1.2), one would have to use for it in the Lagrangians four-curl of A_α in its explicit form. Meanwhile the electromagnetic field Lagrangian does not explicitly contain in this formulation any derivatives of the four-potential. Thus the

Palatini approach ascribes to the expression of the electromagnetic field tensor (2-form)

$$F = dA \tag{4.1.3}$$

a dynamical sense.

Now Maxwell's equations can be written in several equivalent forms:

$$\left.\begin{array}{ll} \delta F = 4\pi j, & dF = 0; \\ F^{\mu\nu}{}_{;\nu} = -4\pi j^{\mu}, & F^{\mu\nu}_{*}{}_{;\nu} = 0; \\ (\sqrt{-g}F^{\mu\nu})_{,\nu} = -4\pi\sqrt{-g}j^{\mu}, & (\sqrt{-g}F^{\mu\nu}_{*})_{,\nu} = 0; \\ d * F = -4\pi * j, & \delta * F = 0. \end{array}\right\} \tag{4.1.4}$$

Here the left-hand column represents variants of the Maxwell equations proper [let us denote all of them as (4.1.4a)], while the right-hand one, (4.1.4b), expresses identical relations if the connection between F and A was imposed initially ($F = dA$), otherwise this is an independent system of equations. It is worth mentioning that asserting F to be a closed form, is not in general equivalent to considering it as an exact form; hence the theory is ramifying at this point.

If one has assumed (4.1.3), only the equations (4.1.4a) have to be considered, these taking the form

$$\Delta A - d\delta A = 4\pi j, \tag{4.1.5}$$

where it is convenient to impose the Lorenz gauge condition,

$$\delta A = 0. \tag{4.1.6}$$

It is worth reminding that this condition is due to L.V. Lorenz of Copenhagen, Denmark, and not to H.A. Lorentz of Leyden, Holland [see Penrose and Rindler (1983)]. Now the de-Rhamian Δ in the left-hand side includes the Ricci curvature tensor, see (1.2.51).

4.2 The Electromagnetic Stress-Energy Tensor and Its Monad Decomposition

The Noether theorem [see Noether (1918), Mitskievich (1958), Mitskievich (1969), Mitskievich, Yefremov and Nesterov (1985)] yields definition of the energy-momentum tensor density:

$$\mathfrak{T}^{\mu\nu} \equiv \sqrt{-g}T^{\mu\nu} = -2\frac{\delta\mathfrak{L}}{\delta g_{\mu\nu}}. \tag{4.2.1}$$

We substitute here the Lagrangian density (4.1.1) using in it a tensor density of the weight $\frac{1}{2}$, $\gamma^{\sigma\tau} = (-g)^{1/4}g^{\sigma\tau}$:

$$\mathfrak{L}_{em} = -\frac{1}{16\pi}F_{\sigma\tau}F_{\alpha\beta}\gamma^{\sigma\alpha}\gamma^{\tau\beta}.$$

An intermediate step will be

$$\mathfrak{T}_{\text{em}}^{\mu\nu} = \frac{1}{4\pi} F_{\sigma\tau} F_{\alpha\beta} \gamma^{\sigma\alpha} \frac{\partial \gamma^{\tau\beta}}{\partial g_{\mu\nu}},$$

where

$$\frac{\partial \gamma^{\tau\beta}}{\partial g_{\mu\nu}} = -(-g)^{1/4}\left(g^{\mu(\tau} g^{\beta)\nu} - \frac{1}{4} g^{\mu\nu} g^{\tau\beta}\right).$$

Remark that $g_{\mu\nu} \frac{\partial \gamma^{\tau\beta}}{\partial g_{\mu\nu}} \equiv 0$ which results in vanishing of trace of the tensor T_{em}. Finally we come to the standard expression

$$T_{\text{em}\,\mu}^{\ \nu} = -\frac{1}{4\pi}\left(F_{\mu\lambda} F^{\nu\lambda} - \frac{1}{4} \delta_\mu^\nu F_{\kappa\lambda} F^{\kappa\lambda}\right), \tag{4.2.2}$$

which can be written in a more symmetric form with help of the "crafty identities" (1.2.6):

$$T_{\text{em}\,\mu}^{\ \nu} = -\frac{1}{8\pi}(F_{\mu\lambda} F^{\nu\lambda} + F^*_{\mu\lambda} F^{\nu\lambda}_*). \tag{4.2.3}$$

Due to (3.1.22), this expression is equivalent to

$$T_{\text{em}\,\mu}^{\ \nu} = -\frac{1}{8\pi}\left(\overset{+}{F}_{\mu\lambda}\overset{-}{F}^{\nu\lambda} + \overset{-}{F}^*_{\mu\lambda}\overset{+}{F}^{\nu\lambda}_*\right) \equiv -\frac{1}{4\pi}\overset{+}{F}_{\mu\lambda}\overset{-}{F}^{\nu\lambda}, \tag{4.2.4}$$

which immediately yields $T_{\text{em}\,\nu}^{\ \nu} \equiv 0$.

The following identities [*cf.* Wheeler (1962)] hold for the electromagnetic stress-energy tensor:

$$T_{\text{em}\,\mu}^{\ \nu} T_{\text{em}\,\nu}^{\ \lambda} = (8\pi)^{-2}[(B^2 - E^2)^2 + 4(E \bullet B)^2]\delta_m^\lambda u, \tag{4.2.5}$$

or equivalently

$$T_{\text{em}\,\mu}^{\ \nu} T_{\text{em}\,\nu}^{\ \lambda} = (8\pi)^{-2}[(B^2 + E^2)^2 - 4(E \times B)^2]\delta_m^\lambda u, \tag{4.2.6}$$

In order to express the electromagnetic stress-energy tensor through the observables, let us consider 1-form $T_{\text{em}\,\mu}^{\ \nu} \tau_\nu dx^\mu$ (the electromagnetic stresses are to be considered separately). Then, making use of (3.1.11) and (3.1.16), we obtain

$$T_{\text{em}\,\mu}^{\ \nu} \tau_\nu dx^\mu = (8\pi)^{-1}(F_{\mu\lambda} E^\lambda - F^*_{\mu\lambda} B^\lambda) dx^\mu. \tag{4.2.7}$$

Since

$$F_{\mu\lambda} E^\lambda dx^\mu = *(E \wedge *F) \quad \text{and} \quad F^*_{\mu\lambda} B^\lambda dx^\mu = -*(B \wedge F), \tag{4.2.8}$$

we have

$$T_{\text{em}\,\mu}^{\ \nu} \tau_\nu dx^\mu = (8\pi)^{-1} * (E \wedge *F + B \wedge F). \tag{4.2.9}$$

After substituting here the decomposition (3.1.17) and its dual conjugate (3.1.20) and easily identifying the scalar and vector products [see (2.2.15)], we obtain finally

$$T_{\text{em}\,\mu}^{\ \nu} \tau_\nu dx^\mu = (8\pi)^{-1}[(E^2 + B^2)\tau + 2E \times B]. \tag{4.2.10}$$

It is clear that the component along the physical time of the reference frame (τ), is the electromagnetic energy density, while the part lying in the three-space of the reference frame, the electromagnetic energy flux density (the Poynting vector), or the numerically identical with it (since $c = 1$) electromagnetic momentum density. We see that all these quantities are expressed through the three-dimensional (but nevertheless generally covariant) electromagnetic field observables (the electric field strength E and magnetic displacement B) in the very same way as they are expressed in the Maxwell theory in flat Minkowski space-time. This compels us to conclude that on the algebraic level, the electrodynamics theory has the same formulation in both special and general relativity, neither the gravitational field, nor the non-inertiality of a reference frame giving any essential contribution in its structure.

As to the purely spatial part of the stress-energy tensor, we have to consider it not as an exterior form, but as a tensor,

$$Stress = T_{\text{em}\,\mu}^{\ \ \nu} b^\mu \otimes b_\nu, \tag{4.2.11}$$

where b^μ and b_μ are semi-coordinated spatial bases (2.2.12), this tensor being (for convenience) of a mixed variance (once covariant and once contravariant). A combination of (4.2.11) and (4.2.3) yields

$$Stress = (8\pi)^{-1}(F_{\mu\lambda}b^\mu \otimes F^{\nu\lambda}b_\nu + F^*_{\mu\lambda}b^\mu \otimes F^{\nu\lambda}_* b_\nu). \tag{4.2.12}$$

In order to translate this to the reference frame language, we use a simple identity Something$^\lambda \equiv$ Something$^\kappa \delta^\lambda_\kappa$, with a subsequent insertion of $\delta^\lambda_\kappa = \tau^\lambda \tau_\kappa + b^\lambda_\kappa$, cf. (2.2.1). The terms with τ yield the field vectors E and B via (3.1.11) and (3.1.16), while the terms with b^λ_κ are expressed through the Levi-Cività axial tensor. We use here the expression (3.1.13) as well as its natural counterpart which follows from the similarity of expressions (3.1.17) and (3.1.20). Further the second identity in (2.2.14) yields the final expression for the *Stress*:

$$Stress = (8\pi)^{-1}[(E^2 + B^2)b^\mu \otimes b_\mu - 2(E \otimes E + B \otimes B)], \tag{4.2.13}$$

the same as in the Minkowski space-time [cf. Synge (1965), p. 323].

On the basis of invariants (3.1.18) and (3.1.19), the standard classification of electromagnetic fields follows, consisting of the three types, electric ($F_{\mu\nu}F^{\mu\nu} < 0$), magnetic ($F_{\mu\nu}F^{\mu\nu} > 0$), and null ($F_{\mu\nu}F^{\mu\nu} = 0$) ones, as it was the case for the Minkowski space-time. Like it was the case also in the flat world, such electromagnetic fields do not in general reduce to purely electric or purely magnetic fields (they are in general not pure fields) in no specific reference frame. However, also as in the Minkowski world, the pure field condition is expressed as vanishing of the second invariant of electrodynamics, $F^*_{\mu\nu}F^{\mu\nu} = 0$. When this condition is fulfilled, it becomes possible to find such reference frames in which the only non-zero observables are either the electric field strength (if $F_{\mu\nu}F^{\mu\nu} < 0$), or the magnetic displacement (if $F_{\mu\nu}F^{\mu\nu} > 0$), or we are dealing with a pure radiation-type field (both invariants are equal to zero), though this last case includes also fields which cannot be considered as propagating waves [*e.g.*, some stationary (zero-frequency) fields; *cf.* (Synge 1974)].

It is remarkable that in the last (pure radiation-type) case, an analogue of the Doppler effect holds: It is possible (in non-wave situations too) to completely transform away the electromagnetic field by switching to moving reference frames, although this can be done only asymptotically ($v \to c$).

4.3 Monad Representation of Maxwell's Equations

First let us write down the action of differential operators on E, the frame-spatial orientation of the resulting quantities being conserved, if these quantities are non-scalar ones. The differential operations will be: The Lie derivative,

$$\pounds_\tau E = E_{\sigma;\lambda}\tau^\lambda b^\sigma + E_\lambda D^\lambda_\sigma b^\sigma + E^\lambda A_{\sigma\lambda}b^\sigma, \qquad (4.3.1)$$

divergence,

$$\operatorname{div} E = \delta E - G \bullet E = E^\alpha{}_{;\alpha} - G \bullet E, \qquad (4.3.2)$$

and curl,

$$\operatorname{curl} E = *(\tau \wedge dE). \qquad (4.3.3)$$

These operations act similarly on the vector B.

Consider now the Maxwell equations (4.1.4a), $\delta F = 4\pi j$, taking into account (3.1.17):

$$\delta F \equiv \delta[(E \wedge \tau) + *(B \wedge \tau)] = 4\pi j. \qquad (4.3.4)$$

Differentiation in the first term yields

$$\delta(E \wedge \tau) = E^\alpha{}_{;\alpha}\tau - E_{\sigma;\lambda}\tau^\lambda dx^\sigma + E_\alpha D^\alpha_\beta dx^\beta - D^\alpha_\alpha E + E^\alpha A_{\alpha\beta}dx^\beta$$

$$= (\operatorname{div} E)\tau - \pounds_\tau E + 2E_\alpha D^\alpha_\beta dx^\beta - D^\alpha_\alpha E, \qquad (4.3.5)$$

and in the second term,

$$\delta * (B \wedge \tau) = *d(B \wedge \tau) = \operatorname{curl} B + G \wedge B - (B \bullet \omega)\tau. \qquad (4.3.6)$$

In the right-hand side of (4.3.4) we have to make use of the decomposition of the four-current (6.1.7), $j = \overset{(\tau)}{u} \rho(\tau + v)$, and to introduce the standard notations for charge and three-current densities,

$$\overset{(\tau)}{\rho} = \overset{(\tau)}{u} \rho, \quad \overset{(3)}{j} = \overset{(\tau)}{u} \rho v, \qquad (4.3.7)$$

which are non-invariant under transitions between different reference frames. We obtain then readily the Maxwell equations with sources, the scalar

$$\operatorname{div} E = 4\pi \overset{(\tau)}{\rho} + \omega \bullet B \qquad (4.3.8)$$

and the vector one,

$$\operatorname{curl} B + G \times B = (\pounds_\tau E - 2E_\nu D^\nu_\mu dx^\mu + D^\alpha_\alpha E) + 4\pi \overset{(3)}{j}. \qquad (4.3.9)$$

The Maxwell equations without sources (4.1.4b), $dF = 0$ (or, equivalently, $\delta * F = 0$), split relative to a reference frame in a complete analogy to the equations (4.1.4a), if we take into account that the expression (3.1.21) differs from F (3.1.18) in (4.3.4) by an exchange of E by $(-B)$ and B by E. Hence the scalar equation takes the form

$$\operatorname{div} B = -\omega \bullet E, \qquad (4.3.10)$$

and the vector one,

$$\operatorname{curl} E + G \times E = -(\pounds_\tau B - 2B_\nu D^\nu_\mu dx^\mu + D^\alpha_\alpha B). \qquad (4.3.11)$$

We see that in the presence of a gravitational field, and in a non-inertial reference frame characterized by acceleration, rotation, and rate-of-strain tensor (thus also in the flat Minkowski world in such a reference frame), the Maxwell equations differ from the customary Maxwell's equations written for an inertial reference frame in the flat world, by the presence of additional terms. Such terms can be interpreted as supplementary effective charges (including an effective "magnetic monopole" charge), as well as the corresponding currents, *cf.* (Mitskievich 1969, Mitskievich and Kalev 1975, Mitskievich, Yefremov and Nesterov 1985). This is in a complete agreement with the known theoretical results claiming appearance of effective electric and magnetic (monopole) charges in rotating reference frames in the presence of magnetic and electric fields correspondingly [see (Dehnen, Hönl and Westpfahl 1961) and (Hönl and Soergel-Fabricius 1961) for an approximate approach]. Such charges and currents have of course a purely kinematic nature, though their objective existence is confirmed by the characteristic structure of the corresponding electric and magnetic fields revealed by their action on charged particles when considered in these reference frames. Obvious examples of this interpretation are presented in the following pages where we consider exact solutions of self-consistent systems of gravitational and electromagnetic fields together with charged perfect fluid.

It is however worth considering the analogy between classical mechanics and electrodynamics in more general terms. In the former theory, the electromagnetic forces acting on charged particles, contain characteristics of the latter ones at most algebraically (in the three-dimensional description, electric field even enters the equations of motion as an inhomogeneity). In the field equations the analogous terms are the sources (mostly inhomogeneity, but the presence of a conducting medium is revealed by a term proportional to the electric field, *i.e.* we encounter the same algebraic dependence, now of the field variable). This analogy between forces and sources can be traced more strictly and formally in the Lagrange-Euler form of equations of motion (for both mechanics and field theory): they correspond to differentiation of the Lagrangian with respect to canonical coordinates (in the field case, the electromagnetic four-potential). Moreover, both the forces and the sources stem from the same terms in the interaction Lagrangian density (which becomes the mechanical interaction Lagrangian when point-like sources are considered). In fact, this is intimately related to an obvious generalization of the third Newtonian law of mechanics to include interaction of mechanical particles and fields. Now, in mechanics a non-inertial frame is revealed by appearance of the inertial forces (the rotational case is usually considered: the centrifugal and Coriolis forces). One has also to expect an appearance of some (let us say) 'inertial sources', these however never having been considered in a far-reaching analogy with the inertial forces of mechanics. But after we have come to the equations (4.3.8)–(4.3.11), this analogy became not a mere possibility, but an inevitable necessity.

The Maxwell field equations seem to have a form much distinct from that of the equations of motion in mechanics, but it is now obvious that the terms connected with rotation of the reference frame (the vector ω) in (4.3.8) and (4.3.10), are direct analogs of the Coriolis force in mechanics. Equations (4.3.9) and (4.3.11) contain the kinematic sources due to the translational acceleration and to deformations (dilatation and shear) of the reference frame; similar terms are known (though less discussed) in mechanics too. The problem now is to find out the role and importance of the corresponding non-inertial effects in the field theory, primarily in electrodynamics. The following pages contain some examples of how this problem is to be handled.

4.4 A Charged Fluid Without Electric Field

We consider here exact solutions of the Einstein-Maxwell equations and equations for a charged perfect fluid, which were obtained as a generalization of the Gödel space-time though having a more restricted isometry group (Mitskievich and Tsalakou 1991). These two families of solutions are described by a general metric

$$ds^2 = e^{2\alpha}(dt + fdx)^2 - e^{2\beta}dx^2 - e^{2\gamma}dy^2 - e^{2\delta}dz^2 . \quad (4.4.1)$$

For the first of the families, (A), one has $e^\alpha = a$, $e^{2\beta} = E\exp(2Bz+C)+\lambda z+\nu$, $2\gamma = 2Bz + C$, $\delta = (\gamma - \beta)\ln a$, $f = bz/a^2$, while the invariant energy density of the fluid is

$$\mu = (\kappa a^2)^{-1}[(b^2/a^2 - \kappa a^2 k^2/2\pi)\exp(-2Bz - C) - 3B^2 E],$$

its pressure $p = B^2 E/(\kappa a^2)$, the invariant charge density

$$\rho = -\sqrt{2}bk/(4\pi a^2)\exp(-2Bz - C),$$

where $\lambda = (\kappa A^2 k^2/\pi - b^2/a^2)/2B$, $\nu = (\lambda + D)/2B$, B, C, D and E being integration constants and k a parameter which controls the switching on and off the charge and the electromagnetic field.

For the second family, (B), $2\alpha = 2Bz + C$, $e^{2\beta} = (b/2B)^2 e^{-2\alpha} - \kappa(akz)^2/2\pi + 2Dz + E$, $e^{2\gamma} = F^2$, $\delta = \alpha - \beta + \gamma$, $f = -(b/2B)e^{-2\alpha}$, while the invariant energy density and pressure of the fluid are

$$\mu = (2\kappa F^2)^{-1}e^{-2\alpha}[2BD + \kappa a^2 k^2(1 - 2Bz)/2\pi],$$

$$p = (2\kappa F^2)^{-1} e^{-2\alpha} [2BD + \kappa a^2 k^2 (1 - 2Bz)/2\pi],$$

and the invariant charge density of the fluid

$$\rho = -\sqrt{2} abk (4\pi F^2)^{-1} e^{-3\alpha}.$$

Here B, C, D, E and F are integration constants. Both families of solutions contain also arbitrary constants a and b. For a special choice of the constants, the electromagnetic field is switched off, and the Gödel (1949) cosmological model is recovered.

We shall not discuss here the method of obtaining these solutions, but their hydrodynamical and electromagnetic properties are of considerable interest. It is natural to employ in their study the co-moving (with respect to the fluid) reference frame described by the monad field

$$\tau = u = e^{\alpha}(dt + f dx). \tag{4.4.2}$$

Since the problem is a stationary one, the rate-of-strain tensor vanishes for both families of solutions, but acceleration and rotation in general survive,

$$G = -\alpha' dz, \quad \omega = f' dy. \tag{4.4.3}$$

Substituting into these expressions the functions entering the families (A) and (B), immediately shows for the family (A) that fluid is in a geodesic motion ($G = 0$), while its rotation is homogeneous in the whole of space ($\omega = (b/a^2) dy$ being an exact form), the rotation axis directed along the y coordinate line. The family (A) fluid may be considered then as a superposition of two distinct fluids, a standard electrically neutral Gödel fluid (if no cosmological term is introduced, it is a stiff matter), and a charged non-coherent dust. In contrast to this case, the fluid in the family (B) not only rotates (its rotation is inhomogeneous everywhere, although the rotation axis is still oriented along the y coordinate line), but moreover its particles are accelerated (non-geodesic motion). This is connected with non-homogeneity of the pressure (as opposite to the family (A) case) and by no means with any interaction with the electromagnetic field (*cf.* section 6.1). The acceleration is directed along z axis, and it behaves exponentially with respect to this coordinate.

For both families (A) and (B), the electromagnetic field is described (in the coordinated basis) by the same potential 1-form

$$A = ka(dt + \sqrt{2} z \, dx). \tag{4.4.4}$$

It is remarkable that this field is of magnetic type since

$$F_{\mu\nu} F^{\mu\nu} = 4a^2 k^2 exp(-2\beta - 2\delta) > 0, \tag{4.4.5}$$

and moreover, it belongs to the pure type ($F^*_{\mu\nu} F^{\mu\nu} = 0$); in the reference frame co-moving with the fluid, electric field is absent completely, in spite of the fact that the fluid is electrically charged (and the sign of the charge does not change anywhere). The magnetic field in the reference frame under consideration is

$$B = \sqrt{2} ak \, exp(\gamma - \beta - \delta) dy. \tag{4.4.6}$$

We have thus come to an apparently paradoxical situation in which a charged fluid does not generate (in a co-moving reference frame) any electric field at all, so that on its particles no lines of force of this field do neither begin nor end, although when these families of solutions were obtained, the Maxwell equations with a non-zero four-current density in the right-hand side were considered (and they were satisfied indeed); for some other examples see Islam (1985). It seems that this fluid generates a magnetic field only, and this does not influence the motion of the particles of this fluid, since the Lorentz force is identically equal to zero (in the co-moving reference frame the particles are at rest by definition). The apparent paradox is resolved very simply [see Mitskievich and Tsalakou (1991)] when it is taken into account that Maxwell's equations should be considered here in a rotating reference frame, taking the form (4.3.8)–(4.3.11). Then, a direct substitution of the quantities characterizing the fluid and electromagnetic field, into *e.g.* equation (4.3.8), shows that there occurs at all points an exact compensation of the charge density by the scalar product of the angular velocity of rotation of the reference frame and the magnetic displacement vector B. This justifies vanishing of the left-hand side of (4.3.8) which takes place simply by virtue of absence of the very electric field. Similar considerations hold concerning all other Maxwell's equations in this reference frame.

In this problem, the crucial fact is that we consider an exact solution of a self-consistent system of equations for the fields and fluid, where the reference frame is objectively determined by the very statement of the problem, so that the final result asserting the possibility of a charged fluid which effectively generates no electric field, becomes obvious from the fact of no electromagnetic interaction between the (non-test) particles of this fluid. And the solution is exact one. The particles of this fluid interact with each other only gravitationally (in the solution B, through pressure too). This effect has however a nature much simpler that the general relativistic one: it may arise in special relativistic and even non-relativistic circumstances. In such a highly non-linear theory as general relativity, it is difficult to immediately conclude whether the fields (both gravitational and electromagnetic ones — though Maxwell's equations remain linear in general relativity too) are generated by the system of charges (and masses) under consideration, or there is a superposition of such a generated field with a free one. This is a case when general relativistic and reference frame considerations cannot help, thus it is worth treating this problem on the Minkowski flat space-time background, see Mitskievich (2001). A readily obtainable conclusion is that in fact both fields are present, that generated dynamically by the charged fluid, and a free (sourceless) field which is essentially magnetic and orthogonal to the three-velocity of the fluid particles. It is easy to see that even in the case of slowly moving fluid, *i.e.* when it is non-relativistic (though the electromagnetic field always represents a relativistic object), such systems of charges and electromagnetic fields are realizable for which the Lorentz force completely vanishes. In the particular case of our generalization of Gödel's solution, we would characterize it as a superposition of the charges Gödel and Bonnor–Melvin universes.

4.5 An Einstein-Maxwell Field with Kinematic Magnetic Charges

Let us consider now another example of non-inertial effects in exact solutions of Einstein-Maxwell equations. This will be a vacuum solution without any material carriers of charges (including, of course, magnetic charges, or dynamical monopoles), but in a rotating reference frame there will appear a distribution of purely kinematic monopoles, *i.e.* we shall have $\operatorname{div} B \neq 0$.

To this end we consider the same form of metric as in the previous section, (4.4.1), but employ another Ansatz for the electromagnetic field,

$$A = kt\, dy, \qquad (4.5.1)$$

which automatically leads to satisfaction of Maxwell's equations without sources. Thus we may consider the case of absence of any matter in our new space-time. The electromagnetic invariants are now

and
$$\left. \begin{array}{l} F_{\mu\nu}F^{\mu\nu} = 2ke^{-2\gamma}(f^2 e^{-2\beta} - e^{-2\alpha}), \\[4pt] F^*_{\mu\nu}F^{\mu\nu} = 0 \end{array} \right\} \qquad (4.5.2)$$

so that electric and magnetic fields should be mutually orthogonal. For simplicity it is natural to admit that the first invariant is also equal to zero, *i.e.* we would be dealing with a null electromagnetic field. Thus we postulate

$$f^2 = e^{2(\beta-\alpha)}. \qquad (4.5.3)$$

The corresponding stress-energy tensor is

$$T_{\text{em}} = \frac{k^2}{4\pi} e^{-2(\alpha-\gamma)}(\theta^{(0)} - \theta^{(1)}) \otimes (\theta^{(0)} - \theta^{(1)}); \qquad (4.5.4)$$

although a null tetrad is more appropriate in this case, we shall use the same orthonormal basis as in the previous section.

Since $\mu = 0$ (no perfect fluid is present), we consider the case $\alpha + \beta = 0$ (this is obvious from the 00 and 11 components of Einstein's equations). A suitable choice of the coordinate z yields $\gamma = \delta$, and we come easily to the following solution:

$$e^{2\alpha} = e^{-2\beta} = f^{-1} = \frac{\kappa k^2}{4\pi} z^2 + Az + B, \quad \gamma = \delta = az + b. \qquad (4.5.5)$$

Then the squared interval takes a remarkably simple form (let $a = 0 = b$),

$$ds^2 = e^{2\alpha}dt^2 + 2dt\, dx - dy^2 - dz^2, \qquad (4.5.6)$$

with null x coordinate lines (or, which is here the same, null hypersurfaces $t = $ const), — a special case of the *pp*-waves[see Kramer, Stephani, MacCallum and Herlt (1980), p. 233ff.].

If we take $\tau = \vartheta^{(0)} = e^\alpha dt + e^{-\alpha} dx$, we get

$$G = -k\alpha' e^{-\delta}\vartheta^{(3)}, \quad \omega = f' e^{-\gamma}\vartheta^{(2)}, \quad D = 0, \tag{4.5.7}$$

as in the previous section, cf. (4.4.3), and

$$E = -k e^{-\alpha-\gamma}\vartheta^{(2)}, \quad B = -k e^{-\alpha-\gamma}\vartheta^{(3)}. \tag{4.5.8}$$

Turning now to Maxwell's equations, we find that $\omega \bullet B = 0$, so that $\mathrm{div} E = 0$, but $\omega \bullet E = k f' e^{-\alpha-2\gamma} \neq 0$, hence $\mathrm{div} B \neq 0$ too, and we come to a distribution of effective magnetic charges as kinematic manifestation of the reference frame rotation. The same conclusions immediately follow from the observations concerning orientation and z-dependence of the (co)vectors E and B. As to the observability of such purely classical kinematic monopoles, we insist that it is perfectly possible, although there is still some deficiency in the definition of the physical three-space of a rotating reference frame, since this submanifold is only local (non-holonomic), so that it is impossible to speak unequivocally about integration over its parts (in order to produce electric and magnetic fluxes). But if infinitesimal regions are considered (and for a continuous distribution of magnetic charges this is just the case), it is quite plausible that such kinematic charges have to be as real as, e.g., the Coriolis force is.

Thus we conclude that when experiments on detection of dynamical magnetic monopoles are performed, it is important to clearly distinguish them from kinematic magnetic charges. However when decay processes are considered, no such a problem could arise, though at a classical level influences of both seem to be largely identical (in the kinematic case, the very rotation of the Earth may simulate a presence of monopoles).

Chapter 5

The Einstein Field Equations

5.1 The Four-Dimensional Einstein Equations

Let us begin with some remarks on the structure of four-dimensional (priory to performing the monad splitting) Einstein's equations. In chapter 4 we mentioned the Palatini method when the Maxwell equations were deduced with the use of the electromagnetic Lagrangian. In the gravitation theory, due to the principle of equivalence, there exists only a relative gravitational force acting on test particles, to which a relative field strength corresponds (we would rather prefer to use the expression "gravitational inhomogeneity field", since observable quantities are in any case determined relative to a reference frame). In order to come to this concept, one may use the geodesic deviation equation,

$$\nabla_u \nabla_u w = \mathsf{R}(u.v)u = R^\alpha{}_{\beta\gamma\delta} u^\beta u^\gamma w^\delta \partial_\alpha, \qquad (5.1.1)$$

which is deduced from the assumption of geodesic motion of free test particles, $\nabla_u u = 0$, and includes a vector field w commuting with the four-velocity field u of the particles, so that w describes closeness of world lines in such a time-like congruence (hence, this equation is related to the focusing effects of the motion of particles, and to the concept of space-time singularities). From (5.1.1) it is clear that the role of the gravitational inhomogeneity field is played by the Riemann–Christoffel curvature tensor (this is the only ingredient in the geodesic deviation equation which does not depend on characteristics of motion of the particles themselves). If one applies the electromagnetic Lagrangian (4.1.1) while throwing off a divergence term according to (4.1.2), it is natural to find an analogy between the quantities

$$A \Leftrightarrow g^{\alpha\beta}, \quad \delta F \Leftrightarrow R^\sigma{}_{\alpha\beta\sigma} \equiv R_{\alpha\beta}. \qquad (5.1.2)$$

Thus we come to the gravitational Lagrangian density

$$\mathfrak{L}_g = \frac{\sqrt{-g}}{16\pi\gamma} g^{\alpha\beta} R_{\alpha\beta} \qquad (5.1.3)$$

in full conformity with the electromagnetic theory, while the constant coefficient involves the Newtonian gravitational constant γ (the Einsteinian constant is also widely used, $\kappa =$

$8\pi\gamma$). Every variational method, including that of Palatini (1919) (*cf.* for other approaches (Landau and Lifshitz 1971, Fock 1964) and (DeWitt 1965) where an important concept of the second variational (functional) derivative is also considered), now yields the well known Einstein gravitational field equations,

$$R_{\alpha\beta} - \frac{1}{2}g_{\alpha\beta}R = -\kappa T_{\alpha\beta}. \tag{5.1.4}$$

Hence it is logical to expect that the monad splitting of these equations (supplemented also by the sourceless equations in analogy with the Maxwell theory), should yield scalar and vector field equations which would exhibit some analogy with Maxwell's equations under the similar splitting (there arise also three-tensor gravitational field equations which are of course *sui generis*).

Any physical field has certainly its own characteristic features (otherwise is would not deserve the status of a separate field); this is true with respect to the gravitational field too whose equations have their own peculiarities. We shall disclose the latter ones in course of the following consideration, but the problem is so extensive and complicated that it is worth being a subject of separate study in another publication: here we are interested mainly in applications of the monad formalism.

5.2 Monad Representation of Einstein's Equations

First we consider equations containing no source terms which are similar to Maxwell's equations (4.3.10) and (4.3.11). These turned to be identities when the electric field strength and magnetic displacement were expressed through the four-potential (the reader may do this easily in the monad language as an exercise); the new gravitational equations are identities in the same sense. Remember expressions for divergence and curl of section 2.4, in particular (2.4.6). Inserting them into (2.4.9) and (2.4.11) (the first of these relations is repeatedly used in the second one), we obtain

$$\text{div}\,\omega = G \bullet \omega \tag{5.2.1}$$

and

$$\text{curl}\,G = \pounds_\tau \omega - 2\omega^\alpha D_{\alpha\beta} dx^\beta + \Theta\omega. \tag{5.2.2}$$

When these equations are compared with Maxwell's equations (4.3.10) and (4.3.11), one sees that they are in fact the same up to a substitution of $-G$ instead of E and ω instead of B; this substitution is exactly the same that can be observed in (3.2.14) and the term analogous to $G \times E$ of (4.3.11) naturally vanishes.

As to Einstein's equations proper, they enter the monad constructions *via* $R_{\alpha\beta}\tau^\alpha\tau^\beta$, $R_{\alpha\beta}\tau^\alpha b^\beta$, $R_{\alpha\beta}b^\alpha \otimes b^\beta$, while it is the best to denote $T_{\alpha\beta}\tau^\alpha\tau^\beta =: \epsilon$, $T_{\alpha\beta}\tau^\alpha b^\beta =: \sigma$, $T_{\alpha\beta}b^\alpha \otimes b^\beta =: \eta$, where ϵ is energy density scalar, σ energy flow density (or equivalently, momentum density) covector, and η tensor of the purely spatial stresses (or, momentum flow density), all taken with respect to the reference frame τ [so that for perfect fluid in a co-moving frame, $\tau = u$, one has $\epsilon = \mu$, $\sigma = 0$, $\eta = p(g - u \otimes u)$, *cf.* (6.1.5); for perfect fluid in an arbitrary frame, ϵ, σ, and η can be read off from (6.1.6)]. Then energy-momentum tensor takes the form

$$T = \epsilon\, \tau \otimes \tau + \tau \otimes \sigma + \sigma \otimes \tau + \eta, \tag{5.2.3}$$

while $\tau \cdot \sigma \equiv 0$, so that the contracted energy-momentum tensor is $\mathrm{tr} T = \epsilon + \mathrm{tr}\eta$.

The obvious identity

$$R_{\alpha\beta}\tau^{\alpha}\tau^{\beta} = \left(\tau^{\alpha}{}_{;\alpha;\beta} - \tau^{\alpha}{}_{;\beta;\alpha}\right)\tau^{\beta}$$

can be easily brought into the form

$$R_{\alpha\beta}\tau^{\alpha}\tau^{\beta} = 2\Theta_{,\alpha}\tau^{\alpha} - G^{\alpha}{}_{;\alpha} + D_{\alpha\beta}D^{\alpha\beta} - A_{\alpha\beta}A^{\alpha\beta}, \tag{5.2.4}$$

and it yields finally a scalar equation [its special case for $G = 0$ is the Raychaudhuri equation, see (Ryan and Shepley 1975)]:

$$\mathrm{div}\, G = \frac{\kappa}{2}(\epsilon - \mathrm{tr}\eta) + 2\Theta_{,\alpha}\tau^{\alpha} - G\bullet G - \frac{1}{2}\omega \bullet \omega + D_{\alpha\beta}D^{\alpha\beta}. \tag{5.2.5}$$

Here we used definition of divergence, (2.4.7), and the fact that $A_{\alpha\beta}A^{\alpha\beta} \equiv \omega \bullet \omega/2$. If now the relation $2G^{\alpha}A_{\alpha\beta}dx^{\beta} = \omega \times G$ is inserted into the generalized Codazzi equations (2.4.14), one obtains the vectorial equation

$$\frac{1}{2}\mathrm{curl}\,\omega = -\kappa\sigma + \omega \times G + \left(D^{\mu}_{\alpha;\nu}b^{\nu}_{\mu} - 2\Theta_{,\alpha}\right)b^{\alpha}. \tag{5.2.6}$$

The tensor equation can be obtained without difficulty on the basis of the generalized Gauss equations (2.4.15), using as well the relation

$$R_{\alpha\beta} \equiv R^{\kappa}{}_{\alpha\beta\lambda}\tau^{\lambda}\tau_{\kappa} + R^{\kappa}{}_{\alpha\beta\lambda}b^{\lambda}_{\kappa},$$

where the left-hand side (Ricci tensor) is to be expressed through the energy-momentum tensor (sources of Einstein's equations), while

$$R^{\kappa}{}_{\alpha\beta\lambda}\tau^{\lambda}\tau_{\kappa} = (\tau_{\alpha;\beta;\lambda} - \tau_{\alpha;\lambda;\beta})\tau^{\lambda},$$

so that this purely three-spatial quantity is easily expressible through the reference frame characteristics only:

$$R_{\alpha\beta\gamma\delta}\tau^{\beta}\tau^{\delta} = \frac{1}{2}G_{\mu;\nu}(b^{\mu}_{\alpha}b^{\nu}_{\gamma} + b^{\mu}_{\gamma}b^{\nu}_{\alpha}) - \pounds_{\tau}D_{\gamma\alpha} - G_{\alpha}G_{\alpha} + D_{\gamma\nu}D^{\nu}_{\alpha}$$

$$+ \frac{1}{4}(\omega_{\alpha}\omega_{\gamma} + \omega \bullet \omega\, b_{\alpha\gamma}) + e_{\mu\delta(\alpha}D^{\delta}_{\gamma)}\omega^{\mu}.$$

At the same time, the crafty identities (1.2.9) yield

$$R_{\alpha\beta\gamma\delta}b^{\alpha}_{\kappa}b^{\beta}_{\lambda}b^{\gamma}_{\mu}b^{\delta}_{\nu} = -R_{\alpha\beta\gamma\delta}\tau^{\beta}\tau^{\delta}e_{\kappa\lambda}{}^{\alpha}e_{\mu\nu}{}^{\gamma} - \frac{1}{2}R(b_{\kappa\mu}b_{\lambda\nu} - b_{\kappa\nu}b_{\lambda\mu})$$

$$- R^{\alpha\gamma}e_{\kappa\lambda\alpha}e_{\mu\nu\gamma} + R^\delta_\beta \tau^\beta \tau_\delta (b_{\kappa\mu}b_{\lambda\nu} - b_{\kappa\nu}b_{\lambda\mu}),$$

but this is exactly the quantity which should be substituted into the right-hand part of the Gauss equations (2.4.15), or in their contracted form (2.4.16), which makes no difference whatever since for a three-dimensional case the Riemann–Christoffel and Ricci tensors are equivalent (they both possess the same number of independent non-trivial components). The subsequent calculations represent a mere routine, yielding finally

$$r_{\lambda\mu} = \frac{1}{2}G_{\kappa;\nu}\left(b^\kappa_\lambda b^\nu_\mu + b^\kappa_\mu b^\nu_\lambda\right) - \pounds_\tau D_{\lambda\mu} - G_\lambda G_\mu + 2D_{\lambda\nu}D^\nu_\mu - 2\Theta D_{\lambda\mu}$$
$$+ e_{\beta\delta(\lambda}D^\delta_{\mu)}\omega^\beta + \frac{\kappa}{2}\epsilon b_{\lambda\mu} - \kappa\left(\eta_{\lambda\mu} - \frac{1}{2}b_{\lambda\mu}\mathrm{tr}\eta\right). \quad (5.2.7)$$

This tensor equation can be found in literature [however, in forms differing much one from the other, not only concerning notations; *cf.* (Vladimirov 1982, Zel'manov and Agakov 1989)]. It has no counterpart in Maxwell's electrodynamics indeed, its characteristic feature being presence of Ricci tensor of the three-dimensional space of the reference frame, so that (5.2.7) may be interpreted also as means for obtaining this latter quantity. We shall not discuss here the important role of obtained equations from the point of view of the Cauchy problem approach.

In the dynamical equations which include sources of the field, specific character of gravitation manifests itself more clearly, although there are certain common terms existing in electromagnetic equations too. The most striking feature of (5.2.6) is absence of time derivative of the acceleration vector [*cf.* $\pounds_\tau E$ in (4.3.9)]. Dehnen (1962) succeeded in eliminating this distinction but at the cost of a highly artificial approach, and a coordinate representation was essential [*cf.* also our approach (Mitskievich 1969) which is probably more transparent]. One knows however from the formulation of the Cauchy problem for Einstein's gravitational field that the noted distinction of Einstein's and Maxwell's theories is a crucial one [see *e.g.* (Mitskievich, Yefremov and Nesterov 1985)].

Thus a question arises, how far-reaching can be the natural analogy between gravitation and electromagnetism (without abandoning the standard context of Einstein's theory of gravity)? The study of the equations of motion of electrically charged test particles in chapter 3 has already shown that this analogy is very profound, starting with the four-dimensional form of these equations, (3.1.5), and being revealed more comprehensively in the monad split equations (3.2.12)–(3.2.14). In this connection there was even extinguished the basic distinction of one theory from another, the uniqueness of the gravitational equivalence principle, although the latter is manifested primarily by equations of motion of the particles. In equation (3.1.5) this enters as the repeated participation of the four-velocity (or the absence of a similar repetition for the four-potential). Equations of the very gravitational field are however much more involved than one would expect: if the set of equations, being in fact identities when acceleration, rotation and rate-of-strain are expressed through the monad field, still remains quasi-Maxwellian in its form, the more profound dynamical equations of gravity, (5.2.4) and (5.2.5), differ from the corresponding Maxwell equations, (4.3.8) and (4.3.9), radically enough.

Does there exist any roundabout way which could outflank this difference without revising the already existing theory (for such a revision there exist no reasons at all according to the profound belief of the most of the specialists)? This way really does exist, but it is not as straightforward one as the approach just considered above. This alternative representation of the gravitation theory is based on the Riemann–Christoffel curvature tensor instead of that of Ricci used in Einstein's equations; moreover, field equations now are first-order differential ones with respect to this tensor (third-order to the metric tensor), while the Ricci tensor enters Einstein's equations purely algebraically. Nevertheless, the physical contents of this higher order theory is the same as it was for Einstein's theory. The gravitational field equations take then the form of contracted Bianchi identities with a substitution of the right-hand side of Einstein's equations instead of the Ricci tensor; then the four-dimensional equations (without introduction of any reference frame approach) show analogy with the Maxwell equations:

$$R^\sigma{}_{\tau\mu\nu;\sigma} = \kappa \left[T_{\tau\nu;\mu} - T_{\tau\mu;\nu} - \frac{1}{2}(g_{\tau\nu}T_{,\mu} - g_{\tau\mu}T_{,\nu}) \right] \tag{5.2.8}$$

[old quasi-Maxwellian gravitational equations; the new ones are (5.4.11)]. A detailed study of these equations see in (Mitskievich 1969) where also the variational formalism (with the corresponding Lagrangian density) was proposed. The first attempt of splitting the gravitational inhomogeneity field $R_{\alpha\beta\gamma\delta}$ in a reference frame sense (not exactly the monad one) was made by Matte (1953) and discussed in a modernized form by Zakharov (1972) [see also Zel'manov and Agakov (1989)]. Hereto the notations were introduced,

$$\overline{X}_{\alpha\beta} = R_{\mu\alpha\nu\beta}\tau^\mu\tau^\nu, \quad \overline{Y}_{\alpha\beta} = -R^*_{\mu\alpha\nu\beta}\tau^\mu\tau^\nu, \quad \overline{Z}_{\alpha\beta} = -R^{**}_{\mu\alpha\nu\beta}\tau^\mu\tau^\nu, \tag{5.2.9}$$

which are clearly similar to (3.1.12) and (3.1.17) for the electromagnetic case (though they are quadratic with respect to projections onto τ). Dashes were written here since we shall use below somewhat different quantities X and Y:

$$X_{\alpha\beta} = C_{\mu\alpha\nu\beta}\tau^\mu\tau^\nu, \quad Y_{\alpha\beta} = -C^*_{\mu\alpha\nu\beta}\tau^\mu\tau^\nu. \tag{5.2.10}$$

5.3 The Geodesic Deviation Equation and a New Level of Analogy Between Gravitation and Electromagnetism

As it was pointed out in section 5.1, the existing exact analogy between gravitation and electromagnetism enables one to formulate the gravitational Lagrangian along the same lines as it was for the electromagnetic one. The directing principle was the use of equations of motion of a particle interacting with the corresponding field in order to determine the field strength concept; for gravitational field one has to consider the geodesic deviation equation (5.1.1). Obviously, it is worth making use of this equation also for finding hints of how to introduce analogues of the electric and magnetic field vectors connected with the gravitational inhomogeneity field.

Let us express the full Riemann–Christoffel tensor in (5.1.1) through the Weyl conformal curvature tensor, Ricci tensor and the scalar curvature. Consider now the term $C_{\alpha\beta\gamma\delta}u^\beta u^\gamma w^\delta$ which may be written also as $C_{\alpha\beta\gamma\delta}dx^{\alpha\beta}dx^{\gamma\delta}(\cdot, u, u, w)$. The Weyl tensor may be considered as the representative of a free gravitational field, the other terms entering the Riemann–Christoffel curvature being expressed through the gravitational field sources [*cf.* the ideas of Pirani and Schild (1961) and Wheeler (1993)]. The quantities X and Y which enter the Weyl tensor, are not only symmetric in their indices, but also traceless, so that they are the only irreducible representatives of free (intrinsic) gravitational inhomogeneity field. We apply here the expression (2.2.21) for the basis 2-forms and note that the necessary contractions are

$$(\tau \wedge b^\mu)(\cdot, u) = \frac{1}{2} \overset{(\tau)}{u} (\tau v^\mu - b^\mu), \tag{5.3.1}$$

$$*(\tau \wedge b^\mu)(\cdot, u) = -\frac{1}{2} \overset{(\tau)}{u} v \times b^\mu, \tag{5.3.2}$$

$$*(\tau \wedge b^\mu)(u, w) = -\frac{1}{2} \overset{(\tau)}{u} (v \times w)^\mu. \tag{5.3.3}$$

Recall now that in the electromagnetic case one has

$$F(\cdot, u) = \frac{1}{2} \overset{(\tau)}{u} [(E \bullet v)\tau + E + (v \times B)] \tag{5.3.4}$$

which gives a combination of the Lorentz force with the work of the field *pro* unit time (and unit charge of the particle); the absence of the second vector, w, is here but natural.

In the case of the Weyl tensor we have

$$C_{\alpha\beta\gamma\delta}dx^{\alpha\beta} \otimes dx^{\gamma\delta}(\cdot, u, u, w) = \left(\overset{(\tau)}{u}\right)^2 [(M \bullet v)\tau + M + (v \times N)], \tag{5.3.5}$$

where

$$M := \left[X_{\alpha\beta}\left(\overset{(\tau)}{w} v^\beta - w^\beta\right) + \frac{1}{2}Y_{\alpha\beta}(v \times w)^\beta\right] b^\alpha \tag{5.3.6}$$

and

$$N := \left[-X_{\alpha\beta}(v \times w)^\beta + \frac{1}{2}Y_{\alpha\beta}\left(\overset{(\tau)}{w} v^\beta - w^\beta\right)\right] b^\alpha. \tag{5.3.7}$$

For X and Y see (5.2.10); it is clear that the analogue of $\overline{Z}$ but without a dash would merely coincide with X, so it is superfluous. The formulae (5.3.4) and (5.3.5) are strikingly similar to each other; the factor $\frac{1}{2}$ in (5.3.4) is due to the definition of the 2-form F, and the second power of the standard relativistic factor $\overset{(\tau)}{u}$ results from the repeated appearance of the four-velocity u in the left-hand side of (5.3.5). It is however impossible to completely identify M and N as exact analogues of the electromagnetic quantities E and B since the formers include a dependence on u, the velocity of particle, and on another vector field w (density of world lines of a cloud of test particles), so they cannot be interpreted as pure field quantities. Consider instead an equation generalizing the geodesic deviation equation (5.1.1) to the case of electrically charged test masses (with equal charge/mass ratio e/m),

$$\nabla_u \nabla_u w = \frac{e}{m}(F_{\alpha\beta}u^\beta)_{;\gamma}w^\gamma dx^\alpha + \mathrm{R}(u,w)u. \tag{5.3.8}$$

We see that there exists certain difference in the structure of the electromagnetic and gravitational (curvature) terms: the latter is quadratic in u. When expressed with respect to a reference frame, equation (5.3.8) contains alongside with the term independent of the three-velocity v (quasi-electric vector) and that linear in the velocity (in electrodynamics, the vector product containing the magnetic displacement vector B), also a term quadratic in the three-velocity of the particle, which cannot have any counterpart in the electromagnetic theory. As to the other vector field, w, it is specific for the deviation type equations, and it is present to the same extent also in the electromagnetic part of (5.3.8).

The only reasonable way out of this situation seems to be an introduction of the already well known symmetric three-tensors, X and Y, present in the motion-of-the-particle dependent quantities M and N, (5.3.6) and (5.3.7). The first is clearly most closely related to the electric type and the other to the magnetic type field, but the intrinsic gravitational inhomogeneity field tensor (the Weyl tensor) exhibits certain peculiarities when entering the geodesic deviation equation. It remains to interpret these as manifestations of the higher tensor rank of the gravitational field.

5.4 New Quasi-Maxwellian Equations of the Gravitational Field

Once contracted Bianchi identities (1.2.30) read

$$R^{\nu\lambda}_{*\ \kappa\nu;\lambda} = 0 \tag{5.4.9}$$

or equivalently,

$$R^\alpha{}_{\beta\gamma\delta;\alpha} = R_{\beta\gamma;\delta} - R_{\beta\delta;\gamma}. \tag{5.4.10}$$

When the right-hand side of Einstein's equations is substituted into the right-hand part of (5.4.10), this yields the old quasi-Maxwellian equations (5.2.8). It is however better to express the Riemann–Christoffel tensor in the left-hand side of (5.4.10) [or (5.2.8)] in terms of the Weyl conformal curvature, Ricci tensor and scalar curvature [see (1.2.31)], and further express the latter ones in terms of the stress-energy tensor and its trace using Einstein's equations.

It is easy to find that (5.4.10) and (5.2.8) yield in this way new equations

$$C_{\sigma\tau\mu\nu}{}^{;\sigma} = \frac{\kappa}{2}[T_{\tau\nu;\mu} - T_{\tau\mu;\nu} - \frac{1}{3}(T_{,\mu}g_{\tau\nu} - T_{,\nu}g_{\tau\mu})] \tag{5.4.11}$$

(in fact, one can find these equations already in the book by Eisenhart (1926), though without their physical interpretation). They hold for arbitrary solutions of Einstein's equations; moreover [see Lichnerowicz (1960)], if a solution of Einstein's equations is taken as initial boundary values, it is reproduced in the whole space-time as a solution of equations (5.2.8) [so of course also of (5.4.11); this situation is closely related to the Cauchy problem]. These equations determine (when considered in analogy with Maxwell's equations for the field tensor F) the intrinsic gravitational inhomogeneity field tensor C (the other parts of the Riemann–Christoffel tensor describe another — extrinsic — kind of the gravitational inhomogeneity field which is algebraically reducible to its sources, the stress-energy tensor T).

It is important that non-contracted Bianchi identities yield exactly the same equations (the contraction does not lead to any information loss in this case): they can be written with help of the Weyl tensor, as it already was the case for the contracted identities [cf. (5.4.2) and then (5.4.11)]. Thus the dual conjugated Weyl tensor now enters the left-hand side as $C^*_{\sigma\tau\mu\nu}{}^{;\sigma}$. But $C^*_{\sigma\tau\mu\nu} \equiv C_{\sigma\tau\mu\nu}{}^*$, so that the resulting quasi-Maxwellian equations are nothing else than the equation (5.4.11) with dual conjugation in the indices μ and ν. As a matter of fact, this is equivalent to representation of the Bianchi identities as vanishing of divergence of the right-hand side of (1.2.9).

Now we have to express the left-hand side of (5.4.11) through the quasi-electric and quasi-magnetic space-like symmetric tensors X and Y. To this end we first substitute relation (2.2.21) into the coordinates-free expression of Weyl's tensor and then take account of the definitions (5.2.10). The resulting decomposition is

$$\frac{1}{4} C_{\alpha\beta\gamma\delta} dx^{\alpha\beta} \otimes dx^{\gamma\delta} = X_{\beta\delta}[(\tau \wedge b^\beta) \otimes (\tau \wedge b^\delta) - *(\tau \wedge b^\beta) \otimes *(\tau \wedge b^\delta)]$$

$$+ Y_{\beta\delta}[(\tau \wedge b^\beta) \otimes (\tau \wedge b^\delta) + *(\tau \wedge b^\beta) \otimes *(\tau \wedge b^\delta)]. \tag{5.4.12}$$

An analogy with the definition (3.1.17) is here obvious, although summations over the indices of X and Y components are shared among both two-forms of the tensor (double bivector) basis. The pure component form of the same expression is

$$C_{\alpha\beta\gamma\delta} = X_{\beta\delta}\tau_\alpha\tau_\gamma + X_{\alpha\gamma}\tau_\beta\tau_\delta - X_{\alpha\delta}\tau_\beta\tau_\gamma - X_{\beta\gamma}\tau_\alpha\tau_\delta - X^{\kappa\lambda}E_{\alpha\beta\mu\kappa}E_{\gamma\delta\nu\lambda}\tau^\mu\tau^\nu$$

$$+ Y_{\beta\lambda}E_{\gamma\delta}{}^{\nu\lambda}\tau_\alpha\tau_\nu - Y_{\alpha\lambda}E_{\gamma\delta}{}^{\nu\lambda}\tau_\beta\tau_\nu + Y_{\lambda\delta}E_{\alpha\beta}{}^{\nu\lambda}\tau_\gamma\tau_\nu - Y_{\lambda\gamma}E_{\alpha\beta}{}^{\nu\lambda}\tau_\delta\tau_\nu. \tag{5.4.13}$$

Consider now the left-hand side of the quasi-Maxwellian equations (5.4.11), *i.e.* four-dimensional divergence of the Weyl curvature. A somewhat lengthy calculation yields the following decomposition with respect to the monad τ:

$$C_{\alpha\beta\gamma\delta}{}^{;\alpha} = -\tau_\beta(\tau_\gamma I_\delta - \tau_\delta I_\gamma) - \tau_\beta II^\lambda e_{\gamma\delta\lambda} + \tau_\gamma III_{\beta\delta} - \tau_\delta III_{\beta\gamma} + IV_\beta{}^\lambda e_{\gamma\delta\lambda} \tag{5.4.14}$$

with

$$\left. \begin{array}{ll} III_{\beta\gamma}e^{\beta\gamma\lambda} - II^\lambda = 0, & III_{\alpha\beta}g^{\alpha\beta} = 0, \\ IV_{\beta\gamma}e^{\beta\gamma\lambda} + I^\lambda = 0, & IV_{\alpha\beta}g^{\alpha\beta} = 0 \end{array} \right\} \tag{5.4.15}$$

(consequences of the four-dimensional Ricci identities for C as well as of its tracelessness). Here

$$I_\delta = X_{\alpha\sigma}{}^{;\alpha}b_\delta^\sigma + G^\alpha X_{\alpha\delta} + 3\omega^\alpha Y_{\alpha\delta} + D^{\alpha\nu}Y_\alpha^\lambda e_{\delta\nu\lambda}, \tag{5.4.16}$$

$$II_\delta = Y_{\alpha\sigma}{}^{;\alpha}b_\delta^\sigma + G^\alpha Y_{\alpha\delta} - 3\omega^\alpha X_{\alpha\delta} - D^{\alpha\nu}X_\alpha^\lambda e_{\delta\nu\lambda}, \tag{5.4.17}$$

$$III_{(\beta\delta)} = \pounds_\tau X_{\beta\delta} - \frac{5}{2}(X_\beta^\alpha D_{\alpha\delta} + X_\delta^\alpha D_{\alpha\beta}) + \frac{1}{4}\omega^\gamma(X_\beta^\alpha e_{\alpha\delta\gamma} + X_\delta^\alpha e_{\alpha\beta\gamma})$$

$$+ 4\Theta X_{\beta\delta} + X_{\mu\nu}D^{\mu\nu}b_{\beta\delta} - G^\nu Y_\beta^\lambda e_{\delta\nu\lambda} - G^\nu Y_\delta^\lambda e_{\beta\nu\lambda} + \frac{1}{2}Y^{\kappa\lambda;\alpha}(e_{\alpha\beta\kappa}b_{\lambda\delta} + e_{\alpha\delta\kappa}b_{\lambda\beta}), \tag{5.4.18}$$

$$IV_{(\beta\delta)} = \pounds_\tau Y_{\beta\delta} - \frac{5}{2}(Y_\beta^\alpha D_{\alpha\delta} + Y_\delta^\alpha D_{\alpha\beta}) + \frac{1}{4}\omega^\gamma(Y_\beta^\alpha e_{\alpha\delta\gamma} + Y_\delta^\alpha e_{\alpha\beta\gamma})$$

$$+ 4\Theta Y_{\beta\delta} + Y_{\mu\nu}D^{\mu\nu}b_{\beta\delta} - G^\nu X_\beta^\lambda e_{\delta\nu\lambda} - G^\nu X_\delta^\lambda e_{\beta\nu\lambda} + \frac{1}{2}X^{\kappa\lambda;\alpha}(e_{\alpha\beta\kappa}b_{\lambda\delta} + e_{\alpha\delta\kappa}b_{\lambda\beta}) \tag{5.4.19}$$

(it is clear from (5.4.15) that only the symmetric parts of III and IV are independent constructions). Since both quasi-electric and quasi-magnetic gravitational inhomogeneity fields are two-index tensors and not vectors, as it was the case for electromagnetic field, it is inappropriate to introduce here operations as curl or div (*cf.* section 4.3), as well as to make use of three-dimensional scalar and vector products.

In a vacuum, the field equations of gravitational field (the higher-order counterpart of Einstein's equations) reduce to vanishing of the quantities (5.4.16)–(5.4.19). A straightforward comparison with the corresponding parts of Maxwell's equations (4.3.8)–(4.3.11) shows existence of every parallels to electrodynamics, with natural differences in the coefficients, while the new specific terms in the gravitational field equations are those which cannot in principle arise when three-dimensional representation of fields is realized with vectors (E, B) and not tensors (X, Y). The only exclusion is appearance of terms with ω^γ, the reference frame rotation vector, in (5.4.18) and (5.4.19), but this can be easily understood since this comes from the expression of the time-derivative, *e.g.*

$$\pounds_\tau X_{\beta\delta} = X_{\beta\delta;\alpha}\tau^\alpha + X_{\alpha\delta}\tau^\alpha{}_{;\beta} + X_{\beta\alpha}\tau^\alpha{}_{;\delta},$$

which contains as compared with $\pounds_\tau E_\beta$, an extra term with $\tau^\alpha{}_{;\delta}$, and the complete compensation [as it was in (4.3.5) and (4.3.1)] is now impossible. In order to facilitate such a comparison, we give here expressions for Maxwell's equations of chapter 4 in the component form:

$$E^\alpha{}_{;\alpha} + G^\alpha E_\alpha + \omega^\alpha B_\alpha = 4\pi \overset{(\tau)}{\rho}, \tag{5.4.20}$$

$$B^\alpha{}_{;\alpha} + G^\alpha B_\alpha - \omega^\alpha E_\alpha = 0, \tag{5.4.21}$$

$$\pounds_\tau E_\delta - 2E^\alpha D_{\alpha\delta} + 2\Theta E_\delta + G^\nu B^\lambda e_{\delta\nu\lambda} + B^{\kappa;\alpha}e_{\alpha\delta\kappa} = -4\pi \overset{(3)}{j}_\delta, \tag{5.4.22}$$

$$\pounds_\tau B_\delta - 2B^\alpha D_{\alpha\delta} + 2\Theta B_\delta - G^\nu E^\lambda e_{\delta\nu\lambda} - E^{\kappa;\alpha}e_{\alpha\delta\kappa} = 0. \tag{5.4.23}$$

It is obvious that this representation of the gravitational field equations is completely in line with the form of Maxwell's equations, in contrast with the direct decomposition of Einstein's equations: in particular, the time-derivatives always enter together with the analogues of curls of the counterpart fields (this was exactly the stumbling-block in section 5.2).

As to the right-hand side of the quasi-Maxwellian equations, we shall restrict our consideration to a perfect fluid in a co-moving reference frame, with the energy-momentum tensor (6.1.5). Then the quasi-Maxwellian equations (5.4.11) take form

$$I_\delta = \frac{\kappa}{3}\mu_{,\nu} b_\delta^\nu, \tag{5.4.24}$$

$$II_\delta = \frac{\kappa}{2}(\mu + p)\omega_\delta, \tag{5.4.25}$$

$$III_{(\beta\delta)} = \kappa(\mu + p)\left[\frac{1}{3}\Theta b_{\beta\delta} - \frac{1}{2}D_{\beta\delta}\right], \tag{5.4.26}$$

$$IV_{(\beta\delta)} = 0, \tag{5.4.27}$$

where for the left-hand sides, expressions (5.4.16)–(5.4.19) should be substituted. The right-hand side expressions are now remarkably simple, and it is easy to check that they satisfy conditions (5.4.15) (since in (5.4.24)–(5.4.27) equations of motion (6.1.8) and (6.1.9) were taken into account, they should be used in this check also).

5.5 Examples of Space-Times

5.5.1 Remarks on Classification of Intrinsic Gravitational Fields

Now it is clear that gravitational inhomogeneity fields can be classified in analogy to the very simple classification of electromagnetic fields which divides the latter ones into the fields of electric, magnetic and null types [see *e.g.* Synge (1956)]. Very naturally, this classification is completely in line with the well known algebraic classification of Petrov (1966) [see also (Géhéniau 1957, Pirani 1957, Bel 1959, Debever 1959, Penrose 1960; Kramer, Stephani, MacCallum and Herlt 1980)].

There exist only four independent invariants of the Weyl tensor C:

$$\left. \begin{array}{ll} I_1 = C_{\alpha\beta\gamma\delta}C^{\alpha\beta\gamma\delta}, & I_2 = C^*_{\alpha\beta\gamma\delta}C^{\alpha\beta\gamma\delta}, \\ I_3 = C_{\alpha\beta\gamma\delta}C^{\alpha\beta\epsilon\eta}C_{\epsilon\eta}{}^{\gamma\delta}, & I_4 = C^*_{\alpha\beta\gamma\delta}C^{\alpha\beta\epsilon\eta}C_{\epsilon\eta}{}^{\gamma\delta} \end{array} \right\} \tag{5.5.1}$$

(they can be found as the identically non-vanishing constructions among the fourteen invariants of the Riemann–Christoffel tensor when the latter is substituted by the Weyl conformal curvature tensor, see *e.g.* (Mitskievich, Yefremov and Nesterov 1985), p. 76). When one expresses the Weyl tensor *via* (5.4.12) or (5.4.13), a representation of the invariants in terms of the quasi-electric and quasi-magnetic gravitational fields emerges:

$$\left. \begin{array}{ll} I_1 = 8(X_{\beta\delta}X^{\beta\delta} - Y_{\beta\delta}Y^{\beta\delta}), & I_3 = 16X_\beta^\alpha(X_\gamma^\beta X_\alpha^\gamma - 3Y_\gamma^\beta Y_\alpha^\gamma), \\ I_2 = -16X_{\beta\delta}Y^{\beta\delta}, & I_4 = 16Y_\beta^\alpha(Y_\gamma^\beta Y_\alpha^\gamma - 3X_\gamma^\beta X_\alpha^\gamma) \end{array} \right\}. \tag{5.5.2}$$

Note that these invariants do not change under both independent infinite-parameter groups of transformations, those of systems of coordinates and of reference frames. They are closely connected with (anti-)self-dual fields (*cf.* the electromagnetic case, section 3.1). To this end observe that

$$\overset{\pm}{C}{}^{*}_{\alpha\beta\gamma\delta} = \pm i \, \overset{\pm}{C}_{\alpha\beta\gamma\delta}, \tag{5.5.3}$$

where

$$\overset{\pm}{C}_{\alpha\beta\gamma\delta} := C_{\alpha\beta\gamma\delta} \mp i C^{*}_{\alpha\beta\gamma\delta}. \tag{5.5.4}$$

Then

$$\overset{\pm}{X}_{\alpha\beta} := \overset{\pm}{C}_{\alpha\mu\beta\nu} \tau^{\mu}\tau^{\nu} = X_{\alpha\beta} \pm i Y_{\alpha\beta} \tag{5.5.5}$$

and finally

$$I_3 \mp i I_4 = \overset{\pm}{C}_{\alpha\beta}{}^{\gamma\delta} \overset{\pm}{C}_{\gamma\delta}{}^{\epsilon\eta} \overset{\pm}{C}_{\epsilon\eta}{}^{\alpha\beta} = 16 \, \overset{\pm}{X}{}^{\beta}_{\alpha} \overset{\pm}{X}{}^{\gamma}_{\beta} \overset{\pm}{X}{}^{\alpha}_{\gamma}. \tag{5.5.6}$$

Similarly,

$$I_1 \mp i I_2 = 8 \, \overset{\pm}{X}{}^{\beta}_{\alpha} \overset{\pm}{X}{}^{\alpha}_{\beta}. \tag{5.5.7}$$

Since the quadratic constructions formed exclusively of the real (not (anti-)self-dual) X or Y are positive definite, it is clear that the sign of I_1 determines the electric (plus) or magnetic (minus) type of the gravitational inhomogeneity field, while $I_1 = 0$ corresponds to the semi-null type field. The gravitational inhomogeneity field is purely quasi-electric, purely quasi-magnetic, or null when all the other invariants vanish (in the case of electromagnetic field the situation is more simple since there exist only two corresponding invariants, and vanishing of the only one second invariant provides the condition of a pure field). When we conclude that the field is purely (quasi-)electric (or magnetic), this does not mean that the alternative field is absent identically: it simply can be transformed away by a choice of the reference frame. The gravitational null case has the same property as its electromagnetic counterpart: the null field merely changes its amplitude under transitions between different reference frames (a manifestation of the Doppler effect in the non-wave representation), with a possibility to transform away (at least, locally) the whole of the field in an asymptotical sense (when the velocity of a new reference frame is tending to the speed of light with respect to the initial frame).

This amazingly complete and simple analogy between electromagnetism and gravitation is well known indeed, but its existence is worth being mentioned since it is widely underestimated. We shall not go into more detail concerning the Petrov classification in connection with the properties of the Weyl tensor invariants, as well as the corresponding properties of X and Y: these aspects may be considered as not so closely related to the reference frames problems.

5.5.2 Example of the Taub–NUT Field

We consider here first the Taub–NUT (in fact, especially NUT) field since it incorporates both quasi-electric and quasi-magnetic properties which are already very well known [for references see *e.g.* (Kramer, Stephani, MacCallum and Herlt 1980)]. Its metric reads

$$ds^2 = e^{2\alpha}(dt + 2l\cos\vartheta d\phi)^2 - e^{-2\alpha}dr^2 - (r^2 + l^2)(d\vartheta^2 + \sin^2\vartheta d\phi^2),$$

so that the natural orthogonal tetrad has to be chosen as

$$\left.\begin{array}{ll} \theta^{(0)} = e^{\alpha}(dt + 2l\cos\vartheta d\phi), & \theta^{(1)} = e^{-\alpha}dr, \\ \theta^{(2)} = (r^2 + l^2)^{1/2}d\vartheta, & \theta^{(3)} = (r^2 + l^2)^{1/2}\sin\vartheta d\phi \end{array}\right\}. \qquad (5.5.8)$$

Here $e^{2\alpha} = (r^2 - 2mr - l^2)/(r^2 + l^2)$. The only non-vanishing independent curvature components in this basis are

$$R_{(0)(1)(0)(1)} = -R_{(2)(3)(2)(3)} = 2U,$$

$$R_{(0)(2)(0)(2)} = R_{(0)(3)(0)(3)} = -R_{(1)(2)(1)(2)} = -R_{(1)(3)(1)(3)} = U,$$

$$R_{(0)(1)(2)(3)} = -2R_{(0)(3)(1)(2)} = -2R_{(0)(2)(3)(1)} = 2V,$$

where

$$\left.\begin{array}{l} U = (l^4 + 3ml^2 r - 3l^2 r^2 - mr^3)(r^2 + l^2)^{-3}, \\ V = l(ml^2 - 3l^2 r - 3mr^2 + r^3)(r^2 + l^2)^{-2} \end{array}\right\}. \qquad (5.5.9)$$

The most convenient choice of the monad field is

$$\tau = \theta^{(0)} = e^{\alpha}(dt + 2l\cos\vartheta d\phi) \qquad (5.5.10)$$

which describes an accelerated and rotating (but without deformations) reference frame,

$$G = -\alpha' e^{\alpha}\theta^{(1)}, \quad \omega = -2le^{\alpha}(r^2 + l^2)^{-1}\theta^{(1)} \qquad (5.5.11)$$

(both vectors have radial orientation, and an effective "spherical symmetry" takes place). Thus from the point of view of section 5.2, we can treat the NUT field as possessing analogues of both electric (G) and magnetic (ω) fields, their radial orientation corresponding to a similarity to the magnetic monopole field superimposed on a point charge field.

The latter interpretation depends however crucially on choice of the reference frame, since it is always possible to consider a non-rotating frame, *e.g.* with

$$\bar{\tau} = N(dt + 2ld\phi), \qquad (5.5.12)$$

$N^2 = (r^2 - 2mr - l^2)(r^2 + l^2)[(r^2 + l^2)^2 - 4l^2(r^2 - 2mr - l^2)\tan^2(\vartheta/2)]$, thus showing analogy with an electric field only. We do not present here the corresponding acceleration vector which has components along axes r and ϑ, since it is rather unsightly.

In the rotating reference frame (5.5.10), the tensors of quasi-electric (gravitoelectric) and quasi-magnetic (gravitomagnetic) gravitational inhomogeneity fields (5.2.9) are diagonal with the components

$$\left.\begin{array}{ll} X_{(1)(1)} = -2U, & X_{(2)(2)} = X_{(3)(3)} = U, \\ Y_{(1)(1)} = 2V, & Y_{(2)(2)} = Y_{(3)(3)} = -V \end{array}\right\}. \qquad (5.5.13)$$

Then the Weyl tensor invariants read

$$I_1 = 48(U^2 - V^2), \quad I_2 = 96UV, \\ I_3 = 96U(U^2 - 3V^2), \quad I_4 = -96V(V^2 - 3U^2) \Bigg\}. \tag{5.5.14}$$

Here the above remarks on the classification of gravitational inhomogeneity fields are clearly applicable.

For the (anti-)self-dual Weyl tensor (5.5.4) and corresponding space-like complex tensors (5.5.5), it is easy to find that

$$U \pm iV = -(r^2 + l^2)^{-3}(m \mp il)(r \pm il)^3 \tag{5.5.15}$$

(*cf.* a superposition of electric point charge and magnetic monopole fields in Maxwell's theory). Then

$$\overset{\pm}{X} = \begin{pmatrix} -2 & 0 & 0 \\ 0 & 1 & 0 \\ 0 & 0 & 1 \end{pmatrix} (U \mp iV). \tag{5.5.16}$$

It is worth mentioning that this type of solution may be obtained in a purely linear theory with the use of cohomology and twistor methods, *cf.* (Hughston 1979).

As to the description of quasi-electric and quasi-magnetic gravitational inhomogeneity fields in a non-rotating reference frame, *e.g.* that of (5.5.12), it involves them both in the Taub–NUT space-time, although they enter not so symmetric as it is the case for the rotating frame [see (5.5.13)]. The transition between the two frames is given by

$$\tau = K\tilde{\tau} + L\tilde{\theta}^{(3)}, \quad \theta^{(3)} = L\tilde{\tau} + K\tilde{\theta}^{(3)},$$

the other two basis covectors remaining unchanged. Here

$$K = [1 - 4l^2 e^{2\alpha}(r^2 + l^2)^{-1} \tan^2(\vartheta/2)]^{-1/1},$$

$$L = 2Kle^{\alpha}(r^2 + l^2)^{-1/2} \tan(\vartheta/2),$$

so that $K^2 - L^2 = 1$, the standard Minkowski relation which holds in general relativity for an orthonormal basis. Now,

$$\tilde{X}_{(1)(1)} = -(2K^2 + L^2)U, \quad \tilde{X}_{(2)(2)} = (K^2 + 2L^2)U, \quad \tilde{X}_{(3)(3)} = U,$$

$$\tilde{Y}_{(1)(1)} = (2K^2 + L^2)V, \quad \tilde{Y}_{(2)(2)} = -(K^2 + 2L^2)V, \quad \tilde{Y}_{(3)(3)} = -V,$$

the tensors remaining diagonal and traceless, while neither of the fields may be transformed away by these means. This situation shows of course more parallels with the electromagnetic field case than the description of gravitation using only G, ω, and D (see section 5.2), and the general transformation laws for X and G can be obtained in strict analogy with electromagnetism [see, *e.g.*, Landau and Lifshitz (1971), although in gravitational case the transformation coefficients enter quadratically, as it is above].

5.5.3 Example of the Spinning Pencil-of-Light Field

Under a pencil of light we understand an infinitesimally thin rectilinear energy distribution moving along itself with the velocity of light. Thus no electromagnetic field is to be considered, and all expressions will be taken outside the singular line of the source. In general the intensity of the source may vary along the line, these variations propagating with the luminal velocity in the same direction. The first (approximate) study of such a situation was undertaken by Tolman, Ehrenfest and Podolsky (1931) [see also (Tolman 1934)]; exact solutions having been found by Peres (1959) and Bonnor (1969) (the latter has considered also the case of a spinning source, and he first gave a proper physical interpretation to the results). Later this subject was revisited by Mitskievich (1981) and Mitskievich and Kumaradtya (1989) who have studied dragging phenomenon in the pencil-of-light field [see also (Mitskievich 1990) and subsection 3.4.3], as well as additivity properties of such fields. From these papers it is clear that the pencil-of-light field possesses both gravitoelectric and gravitomagnetic properties. Formally it may be considered as a gravitational wave field, but, as it is the case in the Einstein–Maxwell pp-wave in section 4.5, this field can be in particular a stationary one, thus being definitively no wave at all.

The simplest description of the spinning pencil-of-light field is connected with a semi-null tetrad,

$$ds^2 = 2\theta^{(0)}\theta^{(1)} - \theta^{(2)}\theta^{(2)} - \theta^{(3)}\theta^{(3)}, \tag{5.5.17}$$

where

$$\left.\begin{array}{ll} \overset{N}{\theta^{(0)}} = dv, & \overset{N}{\theta^{(1)}} = du + Fdv + Hd\phi, \\ \overset{N}{\theta^{(2)}} = d\rho, & \overset{N}{\theta^{(3)}} = \rho d\phi \end{array}\right\}. \tag{5.5.18}$$

Here

$$F = k\ln(\sigma\rho), \quad H = H(v), \tag{5.5.19}$$

H being an arbitrary function of v (as also k and σ may be). The solution is a vacuum one (outside the singular line $\rho = 0$) so that the only non-trivial curvature components

$$\overset{N}{R}_{(0)(2)(0)(2)} = -\overset{N}{R}_{(0)(3)(0)(3)} = \frac{k}{\rho^2}, \quad \overset{N}{R}_{(0)(2)(0)(3)} = -\frac{\dot{H}}{\rho^2}$$

are at the same time components of the Weyl tensor.

For our aims it is better to use an orthonormal tetrad which reads

$$\left.\begin{array}{ll} \theta^{(0)} = 2^{-1/2}[(1+F)dv + du + Hd\phi], & \theta^{(2)} = d\rho, \\ \theta^{(1)} = 2^{-1/2}[(1-F)dv - du - Hd\phi], & \theta^{(3)} = \rho d\phi \end{array}\right\}, \tag{5.5.20}$$

and the non-trivial curvature components are

$$R_{(0)(2)(0)(2)} = R_{(1)(2)(1)(2)} = R_{(0)(2)(1)(2)} = -R_{(0)(3)(0)(3)} = -R_{(1)(3)(1)(3)}$$
$$= -R_{(0)(3)(1)(3)} = k/(2\rho^2),$$

$$R_{(0)(2)(0)(3)} = R_{(1)(2)(1)(3)} = R_{(0)(2)(1)(3)} = R_{(1)(2)(0)(3)} = -\dot{H}/(2\rho^2).$$

Thus in the reference frame with $\tau = \theta^{(0)}$,

$$\left.\begin{array}{l} X(2)(2) = -X(3)(3) = Y(2)(3) = k/(2\rho^2), \\ Y(2)(2) = -Y(3)(3) = -X(2)(3) = \dot{H}/(2\rho^2) \end{array}\right\}, \quad (5.5.21)$$

or for (anti-)self-dual representation,

$$\overset{\pm}{X} = \begin{pmatrix} 0 & 0 & 0 \\ 0 & 1 & \pm 1 \\ 0 & \pm i & -1 \end{pmatrix} (k \pm i\dot{H})/(2\rho^2) \quad (5.5.22)$$

(*cf.* expression (5.5.16) for the NUT field). Here the numerator is an arbitrary complex function of the retarded time v. The field resembles that of a charged rectilinear wire with a current in electrodynamics, but the gravitational case is much richer, incorporating in particular such a property as angular momentum ("polarization") of the source (in electrodynamics this would be probably an effect of the Aharonov–Bohm type).

The monad field $\tau = \theta^{(0)}$ (5.5.20) corresponds not only to an acceleration, but also to a rotation,

$$G = (-k\theta^{(2)} + \dot{H}\theta^{(3)})/(2\rho), \quad \omega = -(\dot{H}\theta^{(2)} + k\theta^{(3)})/(2\rho), \quad (5.5.23)$$

which are here orthogonal to each other. We see that in this representation, as it was the case for (5.5.21) too, both linear and angular momenta of the source (neither of which is transformable away by switching to frames in translational and/or rotational motion), contribute to both quasi-electric and quasi-magnetic parts of the total gravitational field. But in this approach characteristic to section 5.2, the principle of equivalence permits to get rid of rotation or acceleration, and moreover, not only locally, but also in a finite region (we speak then correspondingly on either a normal, or geodesic congruence), so that only in the gravitational inhomogeneity field approach, there is guaranteed the invariant property of a field to be of quasi-electric, quasi-magnetic, or null type, independent of a concrete choice of reference frame. This shows once more that for gravitational fields the feature of inhomogeneity plays a crucial role. We may thus choose a non-rotating monad congruence, e.g. as $\tau = N(dv + du)$, $N^{-2} = 2(1 - k\ln(\sigma\rho)) - (H/\rho)^2$, but this only makes the expressions for X and Y unpleasant, so we shall not consider here this subject in more detail. It is however worth mentioning that a non-rotating frame can be simultaneously introduced in the space-time of a pencil of light only in a limited band of the radial coordinate ρ. This is the so called local stationarity property known also for the ergosphere region of a Kerr black hole [*cf.* Hawking and Ellis (1973)]. This limited band can be moved arbitrarily, but it cannot be expanded to fill the whole space-time of a pencil of light at once.

5.5.4 Gravitational Fields of the Gödel Universe

Finally we come to the Gödel space-time (Gödel 1949) which has certain remarkable properties. We consider its metric in the form

$$ds^2 = a^2[(dt + \sqrt{2}zdx)^2 - z^2 dx^2 - dy^2 - z^{-2} dz^2], \tag{5.5.24}$$

with the corresponding natural orthonormal basis (tetrad) $\theta^{(\alpha)}$.

The Gödel space-time admits five Killing vectors given here [*cf.* Mitskievich (2001)] up to a positive constant factor (say, a^2 occurring in the covariant components):

$$\mathbf{I}: \quad \xi^\mu \partial_\mu = \partial_t, \xi_\mu dx^\mu = dt + \sqrt{2}zdx = a^{-1}\theta^{(0)}; \tag{5.5.25}$$

$$\mathbf{II}: \quad \xi^\mu \partial_\mu = \partial_x, \xi_\mu dx^\mu = \sqrt{2}zdt + z^2 dx; \tag{5.5.26}$$

$$\mathbf{III}: \quad \xi^\mu \partial_\mu = \partial_y, \xi_\mu dx^\mu = -dy; \tag{5.5.27}$$

$$\mathbf{IV}: \quad \xi^\mu \partial_\mu = 2\sqrt{2}z^{-1}\partial_t - 2xz\partial_z - (z^{-2} - x^2)\partial_x,$$
$$\xi_\mu dx^\mu = \sqrt{2}\left(zx^2 + z^{-1}\right)dt + 2z^{-1}xdz + \left(3 + z^2 x^2\right)dx; \tag{5.5.28}$$

$$\mathbf{V}: \quad \xi^\mu \partial_\mu = z\partial_z - x\partial_x, \xi_\mu dx^\mu = -\sqrt{2}zxdt - z^{-1}dz - z^2 xdx. \tag{5.5.29}$$

The Killing vectors **I**, **II** and **IV** are everywhere time-like, **III** space-like, and the Killing vector **V** has a sign-indefinite square, thus in some parts of the space it is time-like and in other parts, space-like. Due to the commutation properties of the three isometries corresponding to **I**, **II** and **IV**, only two of them (always including **I**) can be simultaneously used to introduce coordinates on which the metric coefficients will not depend (the coefficients of their squared differentials will be therefore both positive — time-like axes). The third Killing coordinate (z) available simultaneously with these two, is space-like. Of course, it seems strange to use two time-like axes at once, though this is always exactly the case in dealing with the Gödel space-time. What is not less exotic, there exists the third alternative in introduction of one of this pair of coordinates: it can involve the Killing coordinate corresponding to **V** which excludes the possibility of simultaneous use of the coordinates which are Killingian with respect to **II** and **IV**. The situation is a rare exception in the theory of space-time. It is also possible, using simple transformation of coordinates (in particular, incorporating a into all of them, thus providing them with the usual length dimensionality while the velocity of light remains equal to 1), to bring (5.5.24) to the form

$$ds^2 = e^{2z/a}\left[dt + \sqrt{2}e^{-z/a}(y/a)dz\right]^2 - dx^2 - dy^2 - dz^2 \tag{5.5.30}$$

which in the limit $a \to \infty$ yields the Minkowski space-time.

The co-moving (together with the sources) monad is $\tau = a(dt + \sqrt{2}zdx) = \theta^{(0)}$ which is rotating but geodesic ($G = 0$, $\omega = \sqrt{2}dy$), while the stationarity yields absence of deformations ($D_{\alpha\beta} = 0$). We assume here that the cosmological constant is equal to zero, so that (5.5.24) describes a space-time filled with stiff matter since $\mu = p = (2\kappa a^2)^{-1}$ [see, *e.g.*, (Stephani 1990)]. Then the intrinsic gravitational inhomogeneity field is very simple,

$$-2X_{(1)(1)} = X_{(2)(2)} = -2X_{(3)(3)} = (6a^2)^{-1}, \quad Y_{(\alpha)(\beta)} = 0. \tag{5.5.31}$$

Thus there exists no quasi-magnetic gravitational field in the Gödel universe, although this is rotating (in a sharp contrast with the electromagnetic field in generalizations of this universe where a charged fluid does not produce any electric field in its co-moving frame, see section 4.4).

As to the gravitational inhomogeneity field sources, they may be seen from the quasi-Maxwellian equations (5.4.11) with a substitution of (5.4.16)–(5.4.19) and (5.4.24)–(5.4.27):

$$X_{\alpha\sigma}{}^{;\alpha}b_\delta^\sigma = (\kappa/3)\mu_{,\sigma}b_\delta^\sigma, \tag{5.5.32}$$

$$-3\omega^\alpha X_{\alpha\delta} = \kappa(\mu + p)\omega_\delta, \tag{5.5.33}$$

$$\frac{1}{2}\omega^\gamma(X_\beta^\alpha e_{\alpha\delta\gamma} + X_\delta^\alpha e_{\alpha\beta\gamma}) = 0, \tag{5.5.34}$$

while the equation corresponding to (5.4.19), reduces to $0 \equiv 0$. Equations (5.5.32)–(5.5.34) are satisfied also indeed. We have already excluded from them terms being identically equal to zero, this is the reason why (5.5.33) and (5.5.34) are algebraic with respect to X. In fact (5.5.33) contains an analogue of div B in Maxwell's equations, and it shows how the other terms mutually compensate to permit Y to vanish completely. Similarly, (5.5.34) corresponds to Maxwell's equation with curl B.

Chapter 6

Perfect Fluids

6.1 Introductive Remarks

In this chapter, we consider a description of perfect fluids in the field theoretic context. This is not a merely formal translation of the perfect fluids theory into the field theoretic language but quite a general and more fruitful approach to treat perfect fluids which permits to obtain and practically use new analytic relations between Lagrangian-type constructions in solving self-consistent system of dynamical equations involving also Einstein's equations. Since this theory is comparatively lesser covered in the literature, we give here its introductory overview in order to subsequently include application of the reference frames theory to perfect fluids. More detailed treatment of the general theory of perfect fluids in its field theoretical formulation see in (Mitskievich 1999a,b, 2003).

The main idea of this approach is induced by the observation that the simplest two fields (in certain sense analogous to the electromagnetic one) automatically possess the standard form of stress-energy tensor characteristic for a perfect fluid,

$$T^{\mathrm{pf}} = (\mu + p)u \otimes u - pg, \quad (6.1.1)$$

where μ is invariant energy (mass) density of the fluid and p, its invariant pressure ('invariant' in the sense that they are related to the co-moving frame of the fluid); u is the fluid's four-velocity. These 'simplest' fields do not contain a mass term in their equations, and they are described by skew-symmetric potentials (rank r forms) whose exterior differential represents the corresponding field tensor. Thus the connection coefficients do not enter this description. The Lagrangian densities essentially are functions of quadratic invariants of the field tensors. The most characteristic feature of the stress-energy tensor (6.1.1) is that it has one (single, μ, related to a time-like eigenvector) and one (triple, $-p$, related to three space-like eigenvectors: the Pascal property) eigenvalues (the four eigenvectors form an orthonormal basis), but in general the stress-energy tensor trace does not vanish (in contrast to the electromagnetic case). We do not touch on the energy conditions problem; at least, a part of it can be 'settled' by an appropriate redefinition of the cosmological constant to be then extracted from the stress-energy tensor. Neither shall we consider here thermodynamical properties of fluids, — their phenomenological equations of state will be used instead [see (Kramer et al. 1980)], namely the linear equation

$$p = (2k - 1)\mu \qquad (6.1.2)$$

and the polytropic case,

$$p = A\mu^\Pi. \qquad (6.1.3)$$

The constant k (some authors use instead $\gamma/2$) enters the Lagrangian as $L = -\sigma|J|^k$;[1] J is the quadratic invariant of the field tensor (similar to the first invariant of the Maxwell field) and $\sigma > 0$, a (phenomenological, largely settling dimensionality points) constant. In the usual four-dimensional ($D = 3 + 1$) theory, $k = 1/2$ corresponds to incoherent dust ($p = 0$), $k = 2/3$, to incoherent radiation ($p = \mu/3$), and $k = 1$, to stiff matter ($p = \mu$) in which the sound velocity is equal to that of light.

It is remarkable that only ranks $r = 2$ and 3 automatically correspond to (6.1.1), though only the $r = 2$ case leads to the $\mu + p \neq 0$ term in (6.1.1), but the fluid is then non-rotating due to the $r = 2$ field equations; moreover, in this case one comes to a limited class of equations of state. In the pure $r = 3$ case, the $u \otimes u$ term in (6.1.1) is absent ($p = -\mu$), thus reducing the stress-energy tensor to a pure cosmological term, the corresponding field equation naturally yielding $\mu = $ const (no real dynamics). The $r = 3$ field, however, proves to be necessary, alongside with the $r = 2$ one, for description of rotating fluids, as well as of fluids satisfying more complicated equations of state (e.g., the interior Schwarzschild solution). The scalar field case ($r = 0$) does not meet certain indispensable requirements [see (Mitskievich 1999a,b)] and thus should be dropped, though its stress-energy tensor also is in line with (6.1.1). Our conclusions are essentially based on consideration of the stress-energy tensor of r-form fields ($r = 2$ and 3) and their (rather exotic) dynamics, the fact which makes it clear why our conclusions merely partially coincide with those of Weinberg (1996, section 8.8) who has taken into account only the gauge covariance properties; see Mitskievich (1999b).

Let us first recall some traditional (not yet field-theoretic) features of the perfect fluid theory. A medium consisting of particles which form a perfect (possibly, electrically charged) fluid, can be characterized by the energy-momentum tensor (6.1.1) and the four-current density

$$j = \rho u \qquad (6.1.4)$$

(a vector, not a vector density). Here μ and ρ are invariant (or proper: taken in the co-moving reference frame of the medium) densities of energy (mass) and charge of the fluid, and p is its proper pressure (not to be confused with the four-momentum in (3.2.1): we did not change the somewhat contradictory standard notations which in fact do not meet together in a close context). In such a co-moving frame the energy-momentum tensor takes the form (6.1.1)

$$T^{\mathrm{pf}} = \mu u \otimes u - pw; \qquad (6.1.5)$$

---

[1] This form of L is equivalent to the relation (6.1.2).

here the four-velocity u of the fluid plays the role of monad, and $w = g - u \otimes u$ is the corresponding three-metric tensor [signature of w is $(0, -1, -1, -1)$]. Then in an arbitrary reference frame the energy-momentum tensor of a perfect fluid is

$$T^{\text{pf}} = \left[\overset{(\tau)}{u}{}^2 \mu + \left(\overset{(\tau)}{u}{}^2 - 1 \right) p \right] \tau \otimes \tau$$

$$+ \overset{(\tau)}{u}{}^2 (\mu + p)(\tau \otimes v + v \otimes \tau + v \otimes v) - pb. \quad (6.1.6)$$

Here scalar coefficient in the first term represents the energy (mass) density in the reference frame τ, the coefficient in the second term expresses the proportionality of density of the energy flow (the three-momentum density) of the fluid, to the three-velocity of the fluid in this reference frame τ; the last term describes the isotropic part of pressure (and stresses), while the next to the last one, characterizes its anisotropic part which is due to motion of the fluid with respect to the reference frame. It is interesting that invariant quantities, the energy density and the pressure, taken in the co-moving frame of the fluid, both contribute to the energy density, its flow density, and anisotropic part of the pressure, taken in an arbitrary reference frame, while the fluid is nevertheless a Pascal one. By evaluating the involved quantities one has to take into account the fact that the relativistic factor $\overset{(\tau)}{u}$ is always not smaller than unity $\left(\overset{(\tau)}{u} = dt/ds = (1 - v^2)^{-1/2} \geq 1 \right)$. Expression for the non-invariant densities of the charge and three-current are much simpler,

$$j = \overset{(\tau)}{u} \rho(\tau + v). \quad (6.1.7)$$

It is worth emphasizing that the expressions "invariant" and "non-invariant" are both used for scalar quantities depending not on the choice of a system of coordinates, but on the choice of a reference frame only, while speaking on "three-current" and "three-velocity", *etc*, we mean in fact four-vectors, which are lying in the local submanifold orthogonal to the monad τ. One has to remember that no transformations of coordinates can influence the choice of reference frame, while in the same system of coordinates there could be considered simultaneously as many different reference frames as one would desire. Exactly in this sense the energy density (even if non-invariant under transitions between different choices of τ) automatically is a scalar under arbitrary four-dimensional coordinates transformations, though it might describe energy density of the fluid not in the state of rest. We speak here in such a detail about these elementary facts because they are in an acute contradiction with the vulgar and wide-spread understanding of reference frames as exclusively systems of coordinates.

Now let us touch upon a problem of fluid dynamics in the traditional theory related to vanishing of four-divergence of its energy-momentum tensor, $T^{\text{pf}\mu\nu}{}_{;\nu} = 0$. This means, of course, that we consider in this case an electrically neutral fluid, otherwise it would exchange energy-momentum with the electromagnetic field, so that we should take the latter into account. Moreover, let us consider the fluid's behavior in a co-moving frame, $u = \tau$. From (6.1.5) we have[2]

$$\mu_{,\alpha}\tau^\alpha \equiv \pounds_\tau \mu = -2(\mu + p)\Theta, \qquad (6.1.8)$$

$$p_{,\alpha}b_\beta^\alpha \equiv \left(\overset{(b)}{\operatorname{grad}} p\right)_\beta = (\mu + p)G_\beta. \qquad (6.1.9)$$

We see that a non-homogeneity of the pressure produces a force acting on elements of the fluid and thus making their motion non-geodesic. But if the pressure is homogeneous in the co-moving frame of the fluid, its particles do move geodesically [*cf.* (Synge 1960)], as it is the case for spatially homogeneous cosmological models, otherwise the sum of the pressure and energy density should be equal to zero (in fact, in presence of a cosmological term). The last case is related to the properties of the de Sitter universe.

6.2 Rank 2 and 3 Fields

When the action integral of a physical system is invariant under general transformations of the space-time coordinates, the (second) Noether theorem yields definitions and conservation laws of a set of dynamical characteristics of the system. These are, in particular, its (symmetric) stress-energy tensor (1.3.43) [see also (1.3.41)] and (canonical) energy-momentum pseudotensor (1.3.26).[3] Both objects are mutually connected by the well known Belinfante–Rosenfeld relation (see p. 27). This chapter is focused on a study of the stress-energy tensor of the ranks 2 and 3 fields described by skew-symmetric tensor potentials (2- and 3-forms) whose exterior differentials serve as the corresponding field strengths. This approach does not involve the Christoffel symbols, thus representing the simplest scheme which resembles the general relativistic theory of electromagnetic field.

Similar to the case of electromagnetic field with its 1-form potential A and the field tensor $F = dA$, the 2-form potential B yields the 3-form field tensor $G = dB$, while the 3-form potential C yields the 4-form field tensor $W = dC$. We shall speak respectively on the 1-form [(co)vector] field (electromagnetic field), 2- and 3-form fields [which will describe perfect fluids, as it was shown in Mitskievich (1999a,b) and is summarized below]. To these skew-symmetric fields of rank $r = 1, 2,$ and 3, correspond the quadratic invariants

$$\left.\begin{aligned}
I &= *(dA \wedge *dA) = -(1/2)F_{\mu\nu}F^{\mu\nu}; \\
J &= *(dB \wedge *dB) = -(1/3!)G_{\lambda\mu\nu}G^{\lambda\mu\nu} = \tilde{G}_\kappa \tilde{G}^\kappa; \\
K &= *(dC \wedge *dC) = -(1/4!)W_{\kappa\lambda\mu\nu}W^{\kappa\lambda\mu\nu} = \tilde{W}^2,
\end{aligned}\right\} \qquad (6.2.1)$$

---

[2] The rotation does not contribute here since in $T^{\mu\nu}{}_{;\nu} = 0$ it is cancelled out.

[3] The latter is important in establishment of the commutation relations for the creation and annihilation operators (the second quantization procedure), while the former one acts as the source term in Einstein's field equations.

where $*$ before an object is the Hodge star, and the duality relations hold:

$$B^{\mu\nu}* = \frac{1}{2}E^{\mu\nu\alpha\beta}B_{\alpha\beta}, \quad G_{\lambda\mu\nu} = \tilde{G}^{\kappa}E_{\kappa\lambda\mu\nu}, \quad W_{\kappa\lambda\mu\nu} = \tilde{W}E_{\kappa\lambda\mu\nu}, \qquad (6.2.2)$$

while

$$B^{\mu\nu}{}_{*;\nu} = -\tilde{G}^{\mu}. \qquad (6.2.3)$$

Let us introduce also a unit (co)vector u:

$$u_{\alpha} = \tilde{G}_{\alpha}/J^{1/2}. \qquad (6.2.4)$$

We consider here only the case of $J > 0$, thus u is real and time-like. In the case of free fields we shall use the Lagrangians

$$\mathfrak{L}_1 = \sqrt{-g}L_1(I), \quad \mathfrak{L}_2 = \sqrt{-g}L_2(J), \quad \mathfrak{L}_3 = \sqrt{-g}L_3(K) \qquad (6.2.5)$$

($L = L_1 + L_2 + L_3$). This will yield a description of electromagnetic field without sources, non-rotating electrically neutral fluid, and a cosmological term in Einstein's equations. A general description of rotating fluids can be easily recovered when a specific 'interaction' between ($r = 2$ and 3)-fields is introduced (see below), but this 'interaction' will not result in any terms which could be interpreted as sources in the ($r = 2$ and 3)-fields' equations in a sharp contrast with the electromagnetic field case; instead, the additional term in the rank-two field equations will appear (thus we write 'interaction' in quotation marks). It is also worth being mentioned that while in the electromagnetic theory one is dealing with two field invariants (the second of them is pseudo-invariant, an axial scalar), the 2- and 3-form fields possess merely one quadratic invariant, only that which does not involve dual conjugation. Further in this chapter we do not consider the electromagnetic (1-form) field.

The 2-form and 3-form field equations then are

$$\frac{\delta\mathfrak{L}}{\delta B_{\mu\nu}} = 0 = \frac{\delta\mathfrak{L}}{\delta C_{\lambda\mu\nu}}, \qquad (6.2.6)$$

where $\mathfrak{L} = \sqrt{-g}L$, thus for the 2-form we have

$$\left(\sqrt{-g}\frac{dL_2}{dJ}G^{\lambda\mu\nu}\right)_{,\nu} = 0, \quad \left(\sqrt{-g}\frac{dL_3}{dK}W^{\kappa\lambda\mu\nu}\right)_{,\nu} = 0 \qquad (6.2.7)$$

or, equivalently,

$$d\left(J^{1/2}\frac{dL}{dJ}u\right) \equiv d\left(\frac{dL}{dJ}\tilde{G}\right) = 0 \qquad (6.2.8)$$

(u and $\tilde{G}$ are here covectors). This means that the vector field u is non-rotating.[4] In the case of the 3-form field, its equations read

$$\left(\sqrt{-g}\frac{dL}{dK}W^{\kappa\lambda\mu\nu}\right)_{,\nu} = 0 \qquad (6.2.9)$$

---

[4]This means that the fluid itself is non-rotating.

and are equivalent to
$$K^{1/2} \frac{dL}{dK} = \text{const} \qquad (6.2.10)$$

since $\sqrt{-g} E^{\kappa\lambda\mu\nu} = -\epsilon_{\kappa\lambda\mu\nu} = \text{const}$.

Lagrangian densities of the fields under consideration give according to the relations (1.3.41), (1.3.42), and (1.3.43)

$$T_\alpha^\beta = -L\delta_\alpha^\beta - 2\frac{\partial L}{\partial g_{\mu\beta}} g_{\mu\alpha} \equiv -L\delta_\alpha^\beta + 2\frac{\partial L}{\partial g^{\mu\alpha}} g^{\mu\beta}, \qquad (6.2.11)$$

so that, since L depends on the metric tensor only via J and K,

$$T_\alpha^\beta = -L\delta_\alpha^\beta + 2J\frac{\partial L_2}{\partial J}\left(\delta_\alpha^\beta - u_\alpha u^\beta\right) + 2K\frac{\partial L_3}{\partial K}\delta_\alpha^\beta. \qquad (6.2.12)$$

6.3 Free Rank 2 Field

Let us next consider a free rank 2 field (L_1 being a function only of J), thus the stress-energy tensor (6.2.12) reduces to

$$T_\alpha^\beta = \left(2J\frac{dL_1}{dJ} - L_1\right)\delta_\alpha^\beta - 2J\frac{dL_1}{dJ} u_\alpha u^\beta. \qquad (6.3.1)$$

Here, u evidently is eigenvector of the stress-energy tensor:

$$T_\alpha^\beta u^\alpha = -L_1 u^\beta,$$

while any vector orthogonal to u is also eigenvector, this time with the (triple) eigenvalue $2J\frac{dL_1}{dJ} - L_1$. This is exactly the property of the stress-energy tensor of a perfect fluid, the only additional condition being that the vector u should be a real time-like one. This latter property makes it obvious that u should have the physical interpretation of four velocity of fluid, and its realization depends of course on the concrete solution of the 2-form field equations. Thus we come to a conclusion that

$$\mu = -L_1 \text{ and } p = L_1 - 2J\frac{dL_1}{dJ}, \qquad (6.3.2)$$

μ being mass density and p pressure of the fluid in its co-moving frame.[5]

At the same time, the free 2-form field equations (6.2.8) represent an immediate manifestation of the non-rotation property of perfect fluid. In the section 6.5 we shall find out how one has to modify the Lagrangian L_1 in order to describe also rotating fluids in the field theoretic form.

Perfect fluids characterized by (6.1.2), correspond to a homogeneous function of J as the Lagrangian, $L = -\sigma J^k$, $\sigma > 0$. One may similarly treat the polytropic case, (6.1.3),

---

[5]One may, of course, reinterpret this tensor as a sum of the stress-energy tensor proper and (in general) a cosmological term, *cf.* (Stephani 1990), pp. 87 and 289.

though the Lagrangian is now determined only implicitly. We introduce here a notation $L = -\lambda(J)$; then $\mu + p = \lambda + A\lambda^\Pi = 2J\frac{d\lambda}{dJ}$ and

$$J = \exp\left[2\int \frac{d\lambda}{\lambda + A\lambda^\Pi}\right], \qquad (6.3.3)$$

A and Π being constants.

To sketch a convenient approach to description of a non-rotating fluid, let us use a comoving frame in which the local four-velocity is $u^\mu \sim \delta_0^\mu$, and x^0 coordinate lines form a non-rotating congruence. Since $u \cdot u = 1$,

$$u^\mu = \delta_0^\mu/\sqrt{g_{00}}, \quad \tilde{G}^\mu = \Xi\delta_0^\mu, \qquad (6.3.4)$$

Ξ being a function of coordinates. Thus

$$J = \Xi^2 g_{00}, \text{ and } u^\mu = \tilde{G}^\mu/\sqrt{J}. \qquad (6.3.5)$$

To be more concise, we consider here the case of a homogeneous function $L(J) = \sigma J^k$, hence restricting ourselves to (6.1.2). Then $J^{k-1}\tilde{G}_\lambda = \tilde{\Phi}_{,\lambda}$, $\tilde{\Phi}$ being a pseudopotential (axial scalar). Apart from the metric coefficients, there will be only one independent function characterizing the fluid (and its motion); let us choose it as μ. In the scheme outlined above, this function should be related to Ξ, the only independent function involved in the $r = 2$ field (the metric tensor is supposed to be the same in the perfect fluid and $r = 2$ field languages). Clearly, the problem then reduces to determination of the relationship between these two functions. One immediately finds

$$\mu = \sigma J^k, \text{ thus } \Xi = (\mu/\sigma)^{1/2k}/\sqrt{g_{00}}. \qquad (6.3.6)$$

Hence,

$$\tilde{G}^\mu = \frac{1}{\sqrt{g_{00}}}(\mu/\sigma)^{1/2k}\delta_0^\mu \text{ and } \tilde{\Phi}_{,\mu} = k(\mu/\sigma)^{(2k-1)/2k}g_{0\mu}/\sqrt{g_{00}}. \qquad (6.3.7)$$

6.4 Free Rank 3 Field

In this case the Lagrangian depends only on the invariant K; thus

$$T_\alpha^\beta = \left(2K\frac{dL}{dK} - L\right)\delta_\alpha^\beta = -\frac{\Lambda}{\varkappa}\delta_\alpha^\beta, \qquad (6.4.1)$$

$\varkappa$ being Einstein's gravitational constant. This stress-energy tensor is merely proportional to the metric tensor; therefore the coefficient $2K\frac{dL}{dK} - L = -\Lambda/\varkappa$ obviously should be constant. It is trivially constant (and equal to zero) indeed when $L \sim K^{1/2}$, the field components $W^{\kappa\lambda\mu\nu}$ being then arbitrary. Otherwise, it becomes constant (and nonzero) due to the field equations to which vanishing of the stress-energy tensor divergence is equivalent. Indeed, the equations (6.2.9) reduce to (6.2.10), and we see that both cases (when $L \sim K^{1/2}$ and $L \nsim K^{1/2}$) exactly correspond to the above conclusions. In the first case this does not deserve comments, but when $L \nsim K^{1/2}$, the left-hand side expression in (6.2.10) is really a function of K. Hence from (6.2.10) it follows that K itself should be constant. Thus the "cosmological constant" Λ which appears in (6.4.1), is really constant due to the field equations. These equations, in a sharp contrast to the usual equations of mathematical physics, cannot be characterized as hyperbolic ones (or else). Moreover, the case of $L \sim K^{1/2}$ corresponds to vanishing of the cosmological constant, and the field equations do now impose no conditions on K whatsoever — the rank 3 field is then *arbitrary* due to the field equation, a very specific situation in the field theory.

If $L = \sigma K^k$ with a positive constant σ, then $2k < 1$ corresponds to the de Sitter case; $2k = 1$, to the absence of cosmological constant (this is the case of a *phantom* rank 3 field which is completely arbitrary; moreover, it does not produce any stress-energy tensor at all); finally, $2k > 1$ corresponds to the anti-de Sitter case [see for standard definitions (Hawking and Ellis 1973)]. We propose to call the rank 3 field a Machian (cosmological) field (since another — Machian — reason for this will become obvious *via* consideration of rotating fluids below).

6.5 Rotating Fluids

We came to conclusions that the $r = 2$ and $r = 3$ fields have stress-energy tensors possessing eigenvalues typical to perfect fluids: in the free field cases, the $r = 2$ field with the eigenvalues characteristic for a usual isotropic perfect fluid, and the $r = 3$ field, with only one quadruple eigenvalue (thus the stress-energy tensor is proportional to the metric tensor: the cosmological term form). For description of a perfect fluid with the equation of state $p = (2k - 1)\mu$ and a given constant value of k one needs only one function, say, the mass density μ (the metric tensor is considered as already given, and the system of coordinates is supposed to be co-moving with the fluid, thus the four-velocity vector is $u^\mu = (g_{00})^{-1/2} \delta^\mu_0$). It seemed that this situation in all cases fits well for translating into the $r = 2$ field language. But we were confronted with the no rotation condition for perfect fluid when the rank 2 field was considered to be free. In this case the only remedy is to introduce a source-type term in the $r = 2$ field equations. Speaking in terms of the Lagrangian, this means, at least, to include in the Lagrangian a dependence on the rank 2 field potential B.

The simplest way to do this is to introduce in the Lagrangian density dependence on a new invariant $J_1 = -B_{[\kappa\lambda} B_{\mu\nu]} B^{[\kappa\lambda} B^{\mu\nu]}$ which does not spoil the structure of stress-energy tensor, simultaneously yielding a source term (thus permitting to destroy the no rotation property) without changing the divergence term in the $r = 2$ field equations. We shall use below three invariants: the obvious ones, J and K, and the just introduced invariant of the $r = 2$ field *potential*, J_1. One easily finds that

$$B_{[\kappa\lambda} B_{\mu\nu]} = -\frac{2}{4!} B_{\alpha\beta} B \overset{\alpha\beta}{*} E_{\kappa\lambda\mu\nu}, \qquad (6.5.1)$$

where $B\overset{\alpha\beta}{*} := \frac{1}{2}B_{\mu\nu}E^{\alpha\beta\mu\nu}$ (dual conjugation). Thus $J_1^{1/2} = 6^{-1/2}B_{\alpha\beta}B\overset{\alpha\beta}{*}$. In fact, $J_1 = 0$, if B is a simple bivector ($B = a \wedge b$, a and b being 1-forms; only the four-dimensional case to be considered); this corresponds to all types of rotating fluids discussed in existing literature. This *cannot however annul* the expression which this invariant contributes to the $r = 2$ field equations: up to a factor, it is equal to $\partial J_1^{1/2}/\partial B_{\mu\nu} \neq 0$. Thus let the Lagrangian density be

$$\mathfrak{L} = \sqrt{-g}\left(L(J) + M(K)J_1^{1/2}\right). \tag{6.5.2}$$

The $r = 2$ field equations now take the form [*cf.* (6.2.8)]

$$d\left(\frac{dL}{dJ}\tilde{G}\right) = \sqrt{2/3}M(K)B, \tag{6.5.3}$$

which means that introduction of rotation of the fluid destroys the gauge freedom of the $r = 2$ field. In their turn, the $r = 3$ field equations [*cf.* (6.2.9) and (6.2.10)] yield the first integral

$$J_1^{1/2}K^{1/2}\frac{dM}{dK} = \text{const} \equiv 0 \tag{6.5.4}$$

(when $J_1 = 0$, as it was just stated). It is obvious that K (hence, M) *arbitrarily* depends on the space-time coordinates, if only the $r = 3$ field equations are taken into account. Though the $r = 2$ field equations (6.5.3) apparently show that the $\tilde{G}$ congruence should in general be rotating, the $r = 2$ field B is an exact form for solutions with constant $M(K)$, thus its substitution into the left-hand side of (6.5.3) via $\tilde{G}$ leads trivially to vanishing of G (and hence B). Hence in a non-trivial situation the Machian field K [see (6.2.1)] has to be essentially non-constant.

But the complete set of equations contains Einstein's equations as well. One has to consider their sources and the structure of their solutions (some of which fortunately are available) in order to better understand this remarkable situation probably never encountered in theoretical physics before.

The stress-energy tensor corresponding to the new Lagrangian density (6.5.2), is

$$T_\alpha^\beta = \left(-L - MN + 2J\frac{dL}{dJ} + 2KN\frac{dM}{dK} + 2J_1M\frac{dN}{dJ_1}\right)\delta_\alpha^\beta - 2J\frac{dL}{dJ}u_\alpha u^\beta, \tag{6.5.5}$$

where we have used $N(J_1) = J_1^{1/2}$. It is obvious that only the terms involving L and J survive here ($J_1 = 0 = N$). For a perfect fluid with the equation of state $p = (2k-1)\mu$, one finds $L = -\sigma J^k$, thus $T_\alpha^\beta = -2kLu_\alpha u^\beta + (2k-1)L\delta_\alpha^\beta$.

Then one has a translation algorithm between the traditional perfect fluid and $r = 2$ field languages:

$$\left.\begin{array}{l} \mu = -L = \sigma J^k, \quad \tilde{G}^\mu = \Xi\delta_t^\mu, \quad \Xi = \dfrac{1}{\sqrt{g_{00}}}\left(\dfrac{\mu}{\sigma}\right)^{1/(2k)}, \\[2mm] G = dB = \sqrt{3/2}d\left(\dfrac{1}{M(K)}\right) \wedge d\left(\dfrac{dL}{dJ}\tilde{G}\right) \end{array}\right\} \tag{6.5.6}$$

[cf. (6.5.3)]. The function M depends arbitrarily on coordinates; thus one can choose its adequate form using the last relation without coming into contradiction with the dynamical equations.

We see that the Machian field K plays a very special role in description of rotating fluids. This field makes it possible to consider rotation, but its own field equations do not impose any restriction on K. [A similar situation, but without rotation, was observed in the case of the interior Schwarzschild solution in Mitskievich (1999a).] In each case, one has to adjust the K field using the gravitational field solutions, thus from global considerations (this being the final analysis of considerations of the last paragraphs). Together with the fact that the free K field results in introduction of the cosmological constant, these properties of the Machian (cosmological) field recall the ideas of the Mach principle and a practically forgotten hypothesis of Sakurai (1960).

6.6 Special Relativistic Theory

In special relativity, when $g_{\mu\nu} = \eta_{\mu\nu} = \text{diag}(1, -1, -1, -1)$ (in Cartesian coordinates), one does not use Einstein's equations, so that a homogeneous distribution of a perfect fluid in infinite flat space-time becomes admissible. We shall consider here the behaviour of weak perturbations on the background of such a homogeneous field of a non-rotating perfect fluid. Then in the zeroth approximation $\tilde{G}$ coincides with the four-velocity of the fluid, $u = dt$ (in co-moving coordinates; $t = x^0$), $J = 1$ (the background situation).

Now let a perturbation be introduced, thus

$$\tilde{G}^\kappa = \delta^\kappa_t + \delta\tilde{G}^\kappa, \quad J = 1 + 2\delta\tilde{G}^t + \delta\tilde{G}^\kappa \delta\tilde{G}_\kappa. \tag{6.6.1}$$

These relations might be considered as exact ones, though it is easy to see that, if one does not intend to consider the linear approximation only, it would be worth expressing the very $\delta\tilde{G}$ as a series of terms which describe all orders of magnitude of the perturbations. However in the present context this will be of minor importance, and we shall deal with linear terms only. Then

$$L(J) = L(1) + 2\left[\frac{dL}{dJ}\right]_1 \delta\tilde{G}^t + \ldots; \tag{6.6.2}$$

here the points denote higher-order terms. The expression of $L(J)$ is equivalent (up to its sign) to the mass density, but one still has to take into account the field equations (6.2.8). These read, in similar notations,

$$\tilde{\Phi}_{,\kappa} = \left[\frac{dL}{dJ}\right]_1 \delta^t_\kappa + \left[\frac{dL}{dJ}\delta^\lambda_\kappa + 2\frac{d^2L}{dJ^2}\delta^t_\kappa \delta^\lambda_t\right]_1 \delta\tilde{G}_\lambda + \ldots. \tag{6.6.3}$$

The only property which matters in this expression, is its gradient form. We arrive to the following two equations (the Latin indices being three-dimensional),

$$\left(\tilde{\Phi}\right)_{,t,i} = \left(\tilde{\Phi}\right)_{,i,t} \Rightarrow \left[\frac{dL}{dJ} + 2\frac{d^2L}{dJ^2}\right]_1 \left(\delta\tilde{G}_t\right)_{,i} = \left[\frac{dL}{dJ}\right]_1 \left(\delta\tilde{G}_i\right)_{,t} \tag{6.6.4}$$

and
$$\left(\tilde{\Phi}\right)_{,i,j} = \left(\tilde{\Phi}\right)_{,j,i} \Rightarrow \left[\frac{dL}{dJ}\right]_1 \left(\delta\tilde{G}_i\right)_{,j} = \left[\frac{dL}{dJ}\right]_1 \left(\delta\tilde{G}_j\right)_{,i}. \tag{6.6.5}$$

One has to conclude that this set of equations is satisfied if

$$\delta\tilde{G}_i = \left[\frac{dL/dJ + 2d^2L/dJ^2}{dL/dJ}\right]_1 \left(\int \delta\tilde{G}_t dt + \phi(\vec{x})\right)_{,i}, \tag{6.6.6}$$

with two still non-determined functions, $\delta\tilde{G}_t(t, \vec{x})$ and $\phi(\vec{x})$. But we did not yet taken into account that $\delta\tilde{G}$ (as well as $\tilde{G}$) is divergenceless. This actually means that

$$\delta\tilde{G}^t_{,t} = -\delta\tilde{G}^i_{,i} = \delta\tilde{G}_{i,i} = \left[\frac{dL/dJ + 2d^2L/dJ^2}{dL/dJ}\right]_1 \Delta\left(\int \delta\tilde{G}_t dt + \phi(\vec{x})\right),$$

Δ being the Laplacian operator. Differentiating both sides of this relation with respect to $t = x^0$, we find at last

$$\frac{\partial^2 \delta\tilde{G}_t}{\partial t^2} = \left[\frac{dL/dJ + 2d^2L/dJ^2}{dL/dJ}\right]_1 \Delta\delta\tilde{G}_t, \tag{6.6.7}$$

a modification of the D'Alembert equation (involving a velocity different from that of light). Since propagation properties of perturbations of the mass density μ, of the Lagrangian L and of the field component $\tilde{G}_t$ mutually coincide in the first approximation, one has to conclude that the velocity of the low amplitude density waves (sound propagation) in a fluid is equal to

$$c_s = \sqrt{\left[\frac{dL/dJ + 2d^2L/dJ^2}{dL/dJ}\right]_1} \tag{6.6.8}$$

in units of the velocity of light. One has, of course, to remember that in this theory the laws of thermodynamics were used only implicitly (via equations of state). However some important properties of the sound waves already can be seen in this result.

Let us consider first the simplest case which is described by the equation of state (6.1.2). Then $L = -\sigma J^k$, and we have

$$c_s = \sqrt{2k-1}. \tag{6.6.9}$$

When $k = 1/2$, the perturbations do not propagate (in the co-moving frame of the fluid); this is the case of an incoherent dust whose particles interact only gravitationally, *i.e.* do not interact in a theory devoid of gravitation (special relativity). When $k = 1$, we have a stiff matter, in which (as it is well known) sound propagates with the velocity of light, and this is exactly the case in our field theoretical description: $c_s = 1$. When the value of k lies between $1/2$ and 1, we have more or less realistic fluids, the velocity of sound in them being less than that of light. For example, in the case of incoherent radiation $c_s = 1/3$.

Turning to consideration of a polytropic case (6.1.3) and taking into account its field theoretical description (6.3.3), it is easy to find for the sound velocity (6.6.8) the corresponding

form
$$c_s = \sqrt{\left[1 - 2\left(\frac{dJ}{dL}\right)^{-2} \frac{d^2 J}{dL^2}\right]_1} \qquad (6.6.10)$$

or, after a substitution of (6.3.3), exactly the standard expression

$$c_s = \sqrt{2kp/\mu}. \qquad (6.6.11)$$

It is worth stressing that in this section all considerations were only restricted to absence of gravitational field as well as to weak perturbations of the fluid density, but the velocity of propagation of the perturbations may be relativistic one. Thus the standard expression (6.6.11) represents in fact an exact generalization of c_s to the relativistic case; similarly, (6.6.9) gives correct value of the velocity of sound in ultrarelativistic cases important in astrophysical context.

6.7 Additional Remarks

The rank 3 field does not correspond to any real quantum particles, thus these particles could be nothing more than virtual ones. In the classical theory, the rank 3 field with any degree of non-linearity is equivalent to appearance of cosmological constant in Einstein's equations; when the Lagrangian density is proportional to $K^{1/2}$, the cosmological constant vanishes (thus suggesting a new interpretation of the very fact). The global nature of Mach's principle (admittedly related to rotation phenomena) also seems to justify consideration of the rank 3 field on a basis similar to that of the hypothetical fundamental cosmological field proposed by Sakurai (1960).

The rank 2 field describes (sometimes in interaction with the Machian field) perfect fluids. The second quantization of the free rank 3 field yields real quanta, but they have only spin zero: all other particles appear as thoroughly virtual ones.

It is worth mentioning that in the 2+1-dimensional space-time the $r = 1$ field, formerly, the (non-linear) Maxwell one, now describes perfect fluids, while the $r = 2$ field is responsible for the cosmological term in 3D Einstein's equations, see (Mitskievich 2003).

Chapter 7

Canonical Formalism

The canonical formalism in the theory of gravitation as well as electromagnetic and other fields now experiences a second-wind stage of its development. We include in this chapter a summary of its lesser-known aspects in certain sense related to the ideas of reference frames.

7.1 Mechanics *versus* Field Theory

For our intuition, at least at the stage of its formation, mechanical concepts (dealing with motion of particles) appear to be more elementary than field theoretical and quantum mechanical concepts. The latter ones are initially (if not always) formulated with help of the mechanical concepts, the coordinates (positions), velocities, energies, momenta, masses, and other characteristics borrowed initially from classical mechanics (in quantum mechanics, the concept of wave in classical mechanics of continuous media is also used). At the same time the very classical mechanics (all its concepts and laws) represents a consequence of the field theoretical and quantum mechanical laws and concepts. Thus it does retain quite a few fundamental features of these more elementary (though — seemingly — more complicated) theories. This is why it is worth paying attention to fundamental bases of mechanics which determine all its structure; we will recall them to the reader[1]. At the same time, it would be wrong to think that mechanical concepts could be, without any modification, brought into the field theory (or the quantum theory).

The fundamental notion to characterize a physical system in classical mechanics, is a set of generalized coordinates being functions of one independent parameter, time:

$$q^i = q^i(t), \quad i = 1, \ldots, n, \qquad (7.1.1)$$

where n is the number of degrees of freedom of the physical system under consideration.

---
[1]Among standard treatises on classical mechanics written from the viewpoint of the modern theoretical physics, it is worth mentioning the textbooks by Byelen'ky (1964), Goldstein (1980), Landau and Lifshitz (1976) and Leech (1958).

Derivatives of these coordinates with respect to time are the generalized velocity components,

$$\dot{q}^i, \qquad (7.1.2)$$

so that the Lagrangian function of a mechanical system is a function of these $2n$ arguments:

$$L = L(q, \dot{q}). \qquad (7.1.3)$$

Then the variational principle for the action integral

$$S = \int L \, dt \qquad (7.1.4)$$

yields the Lagrange equations

$$\frac{\delta L}{\delta q^i} := \frac{\partial L}{\partial q^i} - \frac{d}{dt}\frac{\partial L}{\partial \dot{q}^i} = 0, \qquad (7.1.5)$$

equations of motion of a mechanical system. The coordinates (7.1.1) are also called *canonical coordinates*, and the quantities conjugated to them (with respect to the Lagrangian) are the components of canonical momentum (or simply, *canonical momenta*; the velocities (7.1.2) will be called canonical ones too, though they are not used in the canonical Hamiltonian approach proper),

$$p_i := \frac{\partial L}{\partial \dot{q}^i}. \qquad (7.1.6)$$

This introduction of canonical variables in mechanics proved to open wide horizons in the development not only of classical, but also of quantum theory.

The Hamilton function (*Hamiltonian*) of a mechanical system,

$$H(q, p) = p_i \dot{q}^i - L, \qquad (7.1.7)$$

(a contraction in i is meant) is a counterpart of the Lagrangian, being a consequence of application to it of the Legendre transformation. This transformation in general serves to change independent variables, in particular if we differentiate H,

$$dH = \sum_i (dp_i \dot{q}_i + p_i d\dot{q}_i) - \frac{\partial L}{\partial q_i} dq_i - \frac{\partial L}{\partial \dot{q}_i} d\dot{q}_i, \qquad (7.1.8)$$

we find that it depends on q and p, since due to the definition of momentum (7.1.6) the terms containing $d\dot{q}_i$ in (7.1.8), identically cancel. Simultaneously we obtain the standard canonical (Hamiltonian) equations of mechanics

$$\dot{p}_i = -\frac{\partial H}{\partial q^i} \quad \text{and} \quad \dot{q}^i = \frac{\partial H}{\partial p_i}. \qquad (7.1.9)$$

They are $2n$ first-order ordinary differential equations (the Lagrangean equations (7.1.5) are second-order ones, and there exist only n of them). It is widely accepted that Hamiltonian in many cases expresses the system's energy; however in relativistic theory energy represents the time-component of the energy-momentum four-vector.[2]

The concept of field formulated in the 19th century by Faraday and Maxwell, was based on introduction of a certain new continuous medium. It could be thought of as, for example, an elastic or liquid body, the number of whose particles in the unit volume grows indefinitely (in real bodies, this number always remains finite, though one has to introduce some fields of interaction of these particles too). In field theory we consider the case of a new type of continuous medium which is not (in the classical sense) an idealization of the real discrete structure of gases and fluids, *etc.*, described in a continuous manner in hydrodynamics and other parts of the theory (see also the description given in chapter 6). However, like in classical mechanics, we can introduce canonical variables in field theory too. Now spatial coordinates themselves bear no information about state of the system which can be described with the use of the deformation tensor, or the rate-of-strain tensor, together with other characteristics of such a continuous medium [a short and elegant description of this approach gives Leech (1958)]. These characteristics in general depend of all four space-time coordinates as independent variables (in mechanics of point-like particles, one has only one such variable, time).

Thus in field theory the role of canonical coordinates is played by components of (in principle, all) fields,

$$A_a = A_a(x). \tag{7.1.10}$$

Here a is a collective index enumerating both tensor and, probably, matrix (say, spinor) components of all potentials. Einstein's summation convention applies to such indices (we use the first letters of the Latin alphabet), if the contrary is not declared explicitly. The arguments of these potentials are all four space-time coordinates (though we shall write simply x). The Greek indices will take four values: 0, 1, 2, 3, and the Latin ones, from i on, only the spatial values 1, 2, 3.

The number of 'canonical velocities' in field theory,

$$A_{a,\alpha}, \tag{7.1.11}$$

is four times greater than that of 'canonical coordinates' A_a (in contrast to the mechanics of pointlike particles). These quantities are used for constructing the *Lagrangian* (to be more precise, the Lagrangian function density) of a system of fields

$$\mathfrak{L} = \mathfrak{L}(A_a; A_{a,\alpha}) \tag{7.1.12}$$

(some-times higher-order derivatives are to be used, as in the case of Hilbert's gravitational Lagrangian: the second-order derivatives of the metric tensor). The variational principle for

---

[2] The latter cannot be written in the field theory as a true vector under general transformations of coordinates [maybe, the bi-metric formalism or Rylov's approach (Rylov 1964) could help in this respect].

the action integral

$$S = \int \mathfrak{L}(dx) \tag{7.1.13}$$

yields the Euler–Lagrange equations

$$\frac{\delta \mathfrak{L}}{\delta A_a} := \frac{\partial \mathfrak{L}}{\partial A_a} - \frac{\partial}{\partial x^\mu} \frac{\partial \mathfrak{L}}{\partial A_{a,\mu}} = 0 \tag{7.1.14}$$

(the dynamical equations of fields[3]).

In field theory the analogue of canonical momentum (7.1.6) is essentially a density-type quantity (this is why one has to use here a Gothic letter)

$$\mathfrak{P}^{a\alpha} := \frac{\partial \mathfrak{L}}{\partial A_{a,\alpha}}, \tag{7.1.15}$$

where the number of non-trivial (identically) components again is greater than that of the canonical coordinates A_a.[4] While in mechanics the Lagrange equations take form of the 2nd Newtonian law

$$\dot{p}_i = \frac{\partial \mathfrak{L}}{\partial q^i}, \tag{7.1.16}$$

in field theory dynamical field equations can be written in an analogous form

$$\mathfrak{P}^{a\alpha}{}_{,\alpha} = \frac{\partial \mathfrak{L}}{\partial A_a}. \tag{7.1.17}$$

An analysis of the existing parallels between mechanics and field theory, shows that the time derivative is changed in field theory to the four-dimensional gradient. But if it is applied to a quantity which already contains [e.g., in its definition, as it is the case for the momenta (7.1.6) ⇒ (7.1.15)] such a modified 'time derivative', there may occur a summation in both four-dimensional indices (a *divergence operator*), as we see in (7.1.17).

Thus we come to an alternative in the definition of the analogue of *Hamiltonian* (density of the function of Hamilton). On one hand, if we proceed on with the assumption that the Hamiltonian in some situations should describe energy (its density), it could be most naturally identified with the energy-momentum pseudotensor (see section 1.3). To this end we shall put *without contraction* in Greek indices [see Mitskievich (1962b)]

$$p_i \dot{q}^i \Rightarrow \mathfrak{P}^{a\alpha} A_{a,\beta}, \tag{7.1.18}$$

to construct the analogue of (7.1.7) as

$$\mathfrak{t}^\alpha_\sigma := \mathfrak{P}^{a\alpha} A_{a,\sigma} - \mathfrak{L} \delta^\alpha_\sigma. \tag{7.1.19}$$

---

[3] However the structural equations, both in the gravitation and electromagnetic theories, can be also obtained *via* the variational principle, if one uses the Palatini approach.

[4] But in general not necessarily four times greater as it was for the canonical velocities: this is due — in this case — to the absence of some of the canonical velocity components in the Lagrangian. This affects the symmetry properties of (7.1.15) usually referred to existence of *constraints* in the corresponding canonical formalism.

This quantity in many respects is analogous indeed to H (e.g., the Legendre transformation); we shall show this in the next section.

On the other hand, there was proposed another general covariant approach to the canonical formulation of field theory (Brezhnev 1964) with a 'Hamiltonian' $\mathfrak{H} := \mathfrak{P}^{a\alpha} A_{a,\alpha} - \mathfrak{L}$ which however was devoid of immediate sense of energy density.

In a contrast to mechanics, field theory seems not to prohibit higher-order derivatives of potentials to be included in the Lagrangian (see however remarks made on p. 126). Usually these are second-order derivatives, and they, as a rule, enter the Lagrangian in such a manner that the field equations would not exceed the second order. Namely this situation is characteristic for gravitational and electromagnetic fields theory in its usual formulation. Then it would be worth considering Lagrangians which depended on up to the second-order derivatives of potentials (Mitskievich 1958), or on all orders of their derivatives in general (Knapecz 1959, 1960). Then the canonical momenta should be written as

$$\mathfrak{P}^{a\alpha} := \frac{\partial \mathfrak{L}}{\partial A_{a,\alpha}} - \frac{\partial}{\partial x^\beta} \frac{\partial \mathfrak{L}}{\partial A_{a,\alpha,\beta}} =: \frac{\delta \mathfrak{L}}{\delta A_{a,\alpha}}$$

or

$$\mathfrak{P}^{a\alpha} \equiv \frac{\delta \mathfrak{L}}{\delta A_{a,\alpha}} := \sum_{k=0}^{\infty} (-1)^k \frac{\partial^k}{\partial x^{\beta_1} ... \partial x^{\beta_k}} \frac{\partial \mathfrak{L}}{\partial A_{a,\alpha,\beta_1,...,\beta_k}},$$

while the field equations retain the form (7.1.17). However, since we are working with the only field which needs the use of second derivatives of potentials (the gravitational one), and even in this case the Palatini approach is used which reduces the theory to that with the first derivatives of enlarged potentials, the approach described in section 1.3 is completely sufficient also in the dynamical equations.[5]

7.2 Canonical Approach to Field Theory

Under the canonical formalism in classical mechanics one understands such a description of mechanical systems (for example, of point-like particles) which uses as physical variables the canonical coordinates (7.1.1) and canonical momenta (7.1.6), together with the dynamical equations for them, (7.1.9). Then the dynamical equations are first-order ones in time derivatives [not the second-order ones, as it was in the case of the Lagrangian formalism, (7.1.5)].

---

[5]In the case of the Hilbertian Lagrangian of gravitation depending also on second derivatives of canonical coordinates, certain complications arise. To avoid them, these second derivatives are usually excluded being incorporated in a divergence term, leaving then the non-covariant Einsteinian Lagrangian with only the first derivatives. But there is another way, the (generalized) Palatini approach, which resolves the problem with more elegance, though it leads (as this did the straightforward, but lengthy consideration of the second derivatives terms in the full Hilbertian Lagrangian) to results different from those which are related with dropping of the divergence term. Usually this is considered as a defect of the Hilbertian Lagrangian approach, but in our opinion it is rather a fault of the Einsteinian Lagrangian whose contribution to the conservation laws should be reassessed.

In the canonical (Hamiltonian) formalism the uppermost value have not only the classical advantages of it (the use of canonical transformations and the Hamilton–Jacobi equation), but also its heuristic role in the passage from classical to quantum theory (in the case of field theoretical approach, to the second quantization). This is why it is worth trying to formulate here the general structure of the classical Hamiltonian formalism for fields, with a special emphasis on the deduction of Poisson brackets.

As a rule, a canonical formulation of field theory is accompanied by singling out a time coordinate in a rough analogy with the classical non-relativistic mechanics. Such an approach is neither full-fledged enough from the point of view of a covariant theory, nor even unavoidable, if one takes into account the possibility of general covariant (but not necessarily tensorial) Hamiltonian formulation of field theory. We shall give below this formulation built in accordance with the mechanics–field theory analogy discussed in section 7.1.

A more profound interpretation of $\mathfrak{t}^\mu_\nu$ can be now found from the same analogy with classical mechanics, if it were shown that it is connected with $\mathfrak{L}$ by an analogue of the Legendre transformation. The differences stem primarily from the fact that instead of one independent variable, time, of the classical mechanics of particles, we have in field theory four such variables, the space-time (geometrical) coordinates[6]. Thus one has to consider integration of $\mathfrak{t}^\alpha_\sigma$ over a (three-dimensional) hypersurface Σ in the four-dimensional space-time,

$$P_\alpha := \int_\Sigma \mathfrak{t}^\mu_\alpha dS_\mu \equiv \int_\Sigma (\mathfrak{P}^{a\mu} A_{a,\alpha} - \mathfrak{L}\delta^\mu_\alpha) dS_\mu, \qquad (7.2.1)$$

which yields now a functional and not a function as the Hamiltonian H was. The role of analogues of q and p here play A_a and $\mathfrak{P}^{a\alpha}$ with respect to which we have now to consider not a differential, but a variation of P_α:

$$\delta P_\alpha = \int_\Sigma dS_\mu \left(A_{a,\alpha} \delta\mathfrak{P}^{a\mu} + \mathfrak{P}^{a\mu} \delta A_{a,\alpha} - \frac{\partial \mathfrak{L}}{\partial A_a} \delta A_a \delta^\mu_\alpha - \frac{\partial \mathfrak{L}}{\partial A_{a,\nu}} \delta A_{a,\nu} \delta^\mu_\alpha \right). \qquad (7.2.2)$$

Here it is taken into account that geometrical coordinates do not suffer variation, like in classical mechanics of point-like particles, the time t. If the field equations are satisfied, i.e. the Lagrangian (density) contains all the information relevant to both the pure field and interaction parts connected with this field component (A_a), we may substitute (7.1.17) into (7.2.2). Then after a simple rearrangement of its terms, this relation takes the form

$$\delta P_\alpha = \int_\Sigma dS_\mu \left(A_{a,\alpha} \delta\mathfrak{P}^{a\mu} - \mathfrak{P}^{a\mu}{}_{,\alpha} \delta A_a + [(\mathfrak{P}^{a\mu}\delta^\nu_\alpha - \mathfrak{P}^{a\nu}\delta^\mu_\alpha)\delta A_a]_{,\nu} \right). \qquad (7.2.3)$$

---

[6]See also — in the introductory section to this chapter — more comments on the nature of differences between mechanics and field theory, in particular, on the number of canonical coordinates (components of the potential in the field theory) versus the number of canonical velocities. Thus geometrical coordinates are radically different from the canonical ones; the term 'canonical velocities' is, of course, foreign to the canonical formalism which uses only canonical coordinates and canonical momenta, the latter term being intrinsic one for this formalism and needs no definition via other quantities, while in the Lagrangean formalism it is introduced as derivative of Lagrangian with respect to canonical velocity.

The construction entering the divergence — $(\mathfrak{P}^{a\mu}\delta_\alpha^\nu - \mathfrak{P}^{a\nu}\delta_\alpha^\mu)\delta A_a$ — is antisymmetric in its upper indices (there are only two free ones of them, μ and ν), so that one can apply to this term Stokes' theorem reducing the corresponding integration to that over the two-dimensional boundary $\partial\Sigma$ of the initial hypersurface of integration, Σ. If the variations of A_a vanish on $\partial\Sigma$, this divergence term should vanish, and we see that the functional (7.2.1), P_α, depends on A_a and $\mathfrak{P}^{a\alpha}$ only, in a complete (though generalized) conformity with the Hamiltonian H in classical mechanics. Hence (7.2.1) involves a transformation (now, in the functional theory sense) analogous to that of Legendre.

For this obvious reason, $\mathfrak{t}_\nu^\mu$ is interpreted as another representative of the energy-momentum distribution (alongside with the symmetric stress-energy tensor $T_{\mu\nu}$), which justifies its name (*canonical energy-momentum pseudotensor*). One of the two indices in $\mathfrak{t}_\nu^\mu$ — the lower one — corresponds to the number of component of the energy-momentum 'vector' emerging from integration in (7.2.1), while this integration over a hypersurface in the four-dimensional world justifies the existence of the upper index. The splitting of the world in a family of space-like hypersurfaces means fixing a 'simultaneity' of the physical situation under consideration, though this simultaneity is not absolute because the choice of these hypersurfaces is to some extent arbitrary: their most important feature is that they should be space-like (the normal vector at each point of all hypersurfaces of the family being directed to the future), without a constructive determination of simultaneity as it was the case in special relativity. But it is only essential that a world line of any particle (including a photon) should not cross any single hypersurface more than once. One may say that the 'bookkeeping' of energy, momentum, spin and the very objects is performed on and labelled to a hypersurface. It is obvious that one should not take the same object, as well as the same characteristics of it, more than once into account. In fact, this is the causality principle, also in the relativistic approach.

The only problem still outside our scope is that of a general reference frame for such a 'bookkeeping'. If the frame is non-rotating, it admits global (in the region where the frame is realized) hypersurfaces of simultaneity (*i.e.*, orthogonal to the congruence of reference). This is a standard ingredient of all canonical descriptions of relativistic systems (particles or fields or both). But if the frame is rotating, it does not correspond to a hypersurface to which the reference congruence lines should be orthogonal: the congruence is not a normal one, it is said. What could be done is such circumstances, if anything at all? We *live* in such a reference frame and do in it all our bookkeeping, so something certainly can be done. A proposal comes to mind: why should one always use a hypersurface whose local elements are orthogonal to the reference congruence?— let the construction be like tile roof, but with infinitesimal roof tiles, so that no particle (including photons) could avoid crossing the tiles. This is, mathematically speaking, a hypersurface with improper time-like vector field instead of its normal one. Some problems will arise with the Gauss and Stokes theorems, but hopefully only technical ones. Thus one should not let this important possibility out of sight.

Returning to the Hamiltonian theory context, we understand that some density has to be integrated over a space-like hypersurface, in fact, four densities to model the integral energy-momentum. In our approach there is also a possibility to construct a general covariant *scalar* result using the vector density $\mathfrak{w}$ (1.3.51) with a correspondingly adjusted congruence ξ which is equivalent to a four-dimensional continuum of available 'components'. There is 'only' one flaw: in a consistent quantum theory which would include also quantum gravity, there should be no means to judge the hypersurface to be locally spacelike, though a kind of 'space-time' probably could survive at the quantum level.

Let us recall definitions of variational derivatives as coefficients before variations of respective variables inside the integral in variation of the corresponding functional. Depending on the form of concrete integrals, different definitions of variational derivatives on manifolds can be used. In our case, when a functional related to a given 'moment of time', is considered, and the integral is taken for this 'moment' spatially over all the physical system, we have to use variational derivatives on a hypersurface. There are several possibilities to introduce such variational derivatives: one may use either the variants with an additional index corresponding to orientation of the normal to hypersurface in every point of it, or the variant without such an index, though nevertheless taking into account normal vector to the hypersurface, as well as their combinations.

From variation of the integral four-Hamiltonian (7.2.3) (dropping the divergence term in it), we have the complete information to find the corresponding variational derivatives. According to the mentioned alternatives, it is possible to write either

$$\frac{\Delta^\alpha P_\beta}{\Delta \mathfrak{P}^{a\gamma}} = A_{a,\beta} \delta^\alpha_\gamma \qquad (7.2.4)$$

and

$$\frac{\Delta^\alpha P_\beta}{\Delta A_a} = -\mathfrak{P}^{a\alpha}{}_{,\beta}, \qquad (7.2.5)$$

or

$$\frac{\Delta^\alpha P_\beta}{\Delta \mathfrak{P}^{a\gamma}} = A_{a,\beta} n^\alpha n_\gamma \qquad (7.2.6)$$

and

$$\frac{\Delta^\alpha P_\beta}{\Delta A_a} = -\mathfrak{P}^{a\gamma}{}_{,\beta} n^\alpha n_\gamma, \qquad (7.2.7)$$

while when the definition (7.3.13) is used, —

$$\frac{\Delta P_\beta}{\Delta \mathfrak{P}^{a\gamma}} = A_{a,\beta} n_\gamma \qquad (7.2.8)$$

and

$$\frac{\Delta P_\beta}{\Delta A_a} = -\mathfrak{P}^{a\alpha}{}_{,\beta} n_\alpha. \qquad (7.2.9)$$

These pairs of symbolic equations[7] obviously are analogous to the canonical (Hamiltonian) equations in mechanics (7.1.9), thus we call them the canonical (generalized) equations in the field theory.

---

[7] In concrete cases of field Lagrangians, including the interaction terms, these equations take more conventional form, and it is possible to choose those which directly correspond to the traditional theory.

A concrete choice between these equations for their actual use, should be taken from the point of view of the close enough analogy between the resulting Poisson brackets in field theory and the well known Poisson brackets in classical mechanics.[8] If we impose the condition that variational derivatives should not depend on the concrete choice of hypersurface (*i.e.* the normal vector components should not participate in them), the first of the alternatives should be chosen, the equations (7.2.4) and (7.2.5). Moreover, it is easy to verify on the basis of the expression for Poisson brackets given in (7.2.15) that the form (7.2.7) does not lead to contradictions only if the function to be differentiated depends on canonical momenta in a combination with the normal vector: $\mathfrak{P}^{a\alpha} n_\alpha$.

In classical mechanics Poisson's brackets are often introduced *via* time derivative of a function of the canonical coordinates and momenta (one may also consider the Lagrange brackets and find Poisson's ones as their inverse). In field theory, when Poisson's brackets are being introduced, we shall base our considerations on differentiation (with respect to the geometric coordinates x^μ) of a function of canonical coordinates and momenta $F(A_a, \mathfrak{P}^{a\alpha})$ too, first making an observation that the symbolical equations (7.2.4) and (7.2.5) are nothing but the *Hamiltonian equations in field theory*, hence they should be used in deduction of Poisson's brackets in the same way as it was done in classical mechanics.

Let us substitute into the expression for derivative of F,

$$\frac{\partial F}{\partial x^\beta} = \frac{\partial F}{\partial A_a} A_{a,\beta} + \frac{\partial F}{\partial \mathfrak{P}^{a\alpha}} \mathfrak{P}^{a\alpha}{}_{,\beta}, \qquad (7.2.10)$$

the derivatives of canonical coordinates and momenta from the Hamiltonian equations (7.2.4) and (7.2.5):

$$\frac{\partial F(x)}{\partial x^\beta} = \int \left[\frac{\partial F(y)}{\partial A_a(y)} \frac{\Delta^\sigma P_\beta}{\Delta \mathfrak{P}^{a\gamma}(y)} \delta^\gamma(x,y) - \frac{\partial F(y)}{\Delta \mathfrak{P}^{a\gamma}(y)} \frac{\Delta^\gamma P_\beta}{\partial A_a(y)} \delta^\sigma(x,y) \right] dS_\sigma^{(y)}. \qquad (7.2.11)$$

Since

$$\frac{\Delta^\gamma F(x)}{\Delta A_a(y)} = \frac{\partial F(y)}{\partial A_a(y)} \delta^\gamma(x,y) \qquad (7.2.12)$$

and

$$\frac{\Delta^\sigma F(x)}{\Delta \mathfrak{P}^{a\gamma}(y)} = \frac{\partial F(y)}{\Delta \mathfrak{P}^{a\gamma}(y)} \delta^\sigma(x,y), \qquad (7.2.13)$$

we immediately come to Poisson's brackets in field theory:

$$\frac{\partial F(x)}{\partial x^\beta} = \{F(x), P_\beta\} := \int \left[\frac{\Delta^\tau F(x)}{\Delta A_a(y)} \frac{\Delta^\sigma P_\beta}{\Delta \mathfrak{P}^{a\tau}(y)} - \frac{\Delta^\sigma F(x)}{\Delta \mathfrak{P}^{a\tau}(y)} \frac{\Delta^\tau P_\beta}{\Delta A_a(y)} \right] dS_\sigma^{(y)}. \qquad (7.2.14)$$

---

[8]We consider here only the simplest case of the so-called 'non-degenerate fields' having the number of (identically) non-vanishing and linearly independent components $\mathfrak{P}^{a\alpha}$ exactly four times as great as the number of the analogous components A_a, (*e.g.*, for fermion fields). Otherwise one has to introduce constraints [see, *e.g.*, Dirac (1964), Faddeev (1968)] which considerably complicate the theoretical analysis. It seems that 'degenerate fields' always need an individual approach to be used. We shall return to them in brief in the end of this section to find some of their common features.

We had to use here an integral expression for Poisson's brackets, since in the future they will be also written for two functionals, thus not only for one function and one functional. Moreover, only the expression just written explicitly displays the properties characteristic for the usual Poisson brackets (for example, the skew symmetry with respect to a permutation of functionals).

Thus in our approach the classical Poisson brackets in field theory take the form

$$\{F, G\} := \int \left[\frac{\Delta^\tau F}{\Delta A_a(y)} \frac{\Delta^\sigma G}{\Delta \mathfrak{P}^{a\tau}(y)} - \frac{\Delta^\sigma F}{\Delta \mathfrak{P}^{a\tau}(y)} \frac{\Delta^\tau G}{\Delta A_a(y)} \right] dS_\sigma^{(y)}, \qquad (7.2.15)$$

where F and G are, in general, functionals. If they are (a particular case) functions, the definitions (7.2.12) and (7.2.13) yield

$$\{F(x), G(x')\} := \left(\frac{\partial F}{\partial A_a} \frac{\partial G}{\partial \mathfrak{P}^{a\tau}} - \frac{\partial F}{\partial \mathfrak{P}^{a\tau}} \frac{\partial G}{\partial A_a} \right) \delta^\tau(x, x'). \qquad (7.2.16)$$

We give now some concrete expressions for Poisson's brackets in field theory in order to see how far the analogy between mechanics and field theory goes in this respect. But first let us introduce some useful notations

$$\mathfrak{P}^a = \int \mathfrak{P}^{a\sigma} dS_\sigma \quad \text{and} \quad A_{a\sigma} = \int A_a dS_\sigma. \qquad (7.2.17)$$

Substituting now concrete quantities in Poisson's brackets, we find

$$\{A_a(x), \mathfrak{P}^{b\alpha}(x')\} = \delta_a^b \delta^\alpha(x, x'), \qquad (7.2.18)$$
$$\{A_a(x), \mathfrak{P}^b\} = \delta_a^b, \qquad (7.2.19)$$
$$\{A_{a\sigma}, \mathfrak{P}^{b\alpha}(x')\} = \delta_a^b \delta_\sigma^\alpha, \qquad (7.2.20)$$
$$\{A_{a\sigma}, \mathfrak{P}^b\} = \delta_a^b \int dS_\sigma, \qquad (7.2.21)$$
$$\{A_a(x), \mathfrak{M}_\beta^{\tau\alpha}(x')\} = A_a|_\beta^\alpha(x) \delta^\tau(x, x'), \qquad (7.2.22)$$
$$\{A_a(x), S_\beta^\alpha\} = A_a|_\beta^\alpha(x), \qquad (7.2.23)$$
$$\{A_{a\sigma}, S_\beta^\alpha\} = \int A_a|_\beta^\alpha(x) dS_\sigma, \qquad (7.2.24)$$
$$\{\mathfrak{M}_\beta^{\tau\alpha}(x), \mathfrak{P}^{a\gamma}(x')\} = \mathfrak{P}^{b\tau} A_b^a|_\beta^\alpha \delta^\gamma(x, x'), \qquad (7.2.25)$$
$$\{S_\beta^\alpha, \mathfrak{P}^{a\gamma}\} = \mathfrak{P}^{b\tau} n_\tau n^\gamma A_b^a|_\beta^\alpha, \qquad (7.2.26)$$
$$\{\mathfrak{M}_\beta^{\tau\alpha}(x), \mathfrak{P}^a\} = \mathfrak{P}^{b\tau}(x) A_b^a|_\beta^\alpha(x), \qquad (7.2.27)$$
$$\{S_\beta^\alpha, \mathfrak{P}^a\} = \int \mathfrak{P}^{b\sigma} A_b^a|_\beta^\alpha dS_\sigma, \qquad (7.2.28)$$
$$\{S_\beta^\alpha, P_\sigma\} = \int \mathfrak{M}_{\beta,\sigma}^{\omega\alpha} dS_\omega, \qquad (7.2.29)$$

which it is easy to verify.

Primarily, from these expressions it is clear that the operation of integration can be inserted inside Poisson's brackets, as if they represented merely algebraic constructions built of the quantities written in these brackets. For example,

$$\int \{\mathfrak{M}_\beta^{\tau\alpha}(x), \mathfrak{P}^{a\gamma}(x')\} dS_\tau^{(x')} = \mathfrak{P}^{b\tau} A_b^a|_\beta^\alpha n_\tau n^\gamma, \tag{7.2.30}$$

is to be compared with (7.2.26) and (7.2.25); or either [see (7.2.29)]

$$\left. \begin{array}{rcl} \{\mathfrak{M}_\beta^{\tau\alpha}, P_\sigma\} & = & \mathfrak{M}_{\beta,\sigma}^{\tau\alpha}, \\ \int \{\mathfrak{M}_\beta^{\tau\alpha}, P_\sigma\} dS_\tau & = & \int \mathfrak{M}_{\beta,\sigma}^{\tau\alpha} dS_\tau \end{array} \right\}. \tag{7.2.31}$$

This degree of consistency makes it most probable that the usual quantization postulate (for Poisson's brackets and commutators or anticommutators, connecting classical particle mechanics with quantum mechanics) could be formulated also in field theory on the basis of the brackets proposed here.

Secondly, it is easy to verify that the classical Poisson brackets in field theory possess the following standard algebraic properties: the already mentioned skew symmetry

$$\{F, G\} = -\{G, F\}; \tag{7.2.32}$$
$$\{F, G + H\} = \{F, G\} + \{F, H\} \text{ (linearity)}; \tag{7.2.33}$$
$$\{F, GH\} = G\{H, G\} + \{F, G\}H \text{ (distributivity)}; \tag{7.2.34}$$
$$\{F, \{G, H\}\} + \{G, \{H, F\}\} + \{H, \{F, G\}\} = 0$$
$$\text{(the Jacobi identity)}; \tag{7.2.35}$$
$$\frac{\partial}{\partial A_a}\{F, G\} = \left\{\frac{\partial F}{\partial A_a}, G\right\} + \left\{F, \frac{\partial G}{\partial A_a}\right\} \tag{7.2.36}$$

[similar to the Leibniz property which can be interpreted also as (7.2.34)]. These properties should be of use in formulation of the basis of the second quantization of fields.

It is worth observing that the Hamiltonian equations (7.2.4) and (7.2.5) follow from the variational principle. To this end let us introduce the integral four-Lagrangian (now, not a density)

$$L_\beta := \int \mathfrak{L} dS_\beta \tag{7.2.37}$$

which is a part of the energy-momentum 'vector'

$$P_\beta = \int \mathfrak{P}^{a\alpha} A_{a,\beta} dS_\alpha - L_\beta. \tag{7.2.38}$$

Extremum of the action integral

$$S = \int L_\beta dx^\beta = \int \mathfrak{P}^{a\alpha} A_{a,\beta} dS_\alpha dx^\beta - \int L_\beta dx^\beta \tag{7.2.39}$$

obtained by a variation with respect to canonical coordinates and momenta, is then expressed via the field equations in the form of (7.2.4) and (7.2.5).

In order to obtain equations generalizing the Hamilton–Jacobi equation of classical mechanics, one has to shift the upper (in the sense of time) limit of the integration region in (7.2.39); it is convenient then to take this region as a hypercylinder. Some calculation yields

$$\frac{\partial S}{\partial x^\mu} = -P_\mu; \tag{7.2.40}$$

since a similar variation leads to

$$\frac{\Delta^\alpha S}{\Delta A_a} = \mathfrak{P}^{a\alpha}, \tag{7.2.41}$$

the equations (7.2.40) can be called the field theoretical symbolical Hamilton–Jacobi equations in which the integral four-Hamiltonian P_μ has to be considered as a functional of A_a and $\Delta^\alpha S/\Delta A_a$. Thus we now have a differential-functional equation as a generalization from particle mechanics to field theory, and not simply a differential equation as it was in the case of mechanics.

In this theory it is possible to introduce canonical transformations for fields (Mitskievich 1965). It is interesting that these transformations are strictly analogous to the canonical (contact) transformations of classical mechanics (when one takes, of course, an integral over a hypersurface). On the other hand, classical field theory may be considered as a theory with 'primary quantization' (but not second quantization) in comparison with mechanics of point-like particles, whereas the standard procedure of this quantization leads to the well known change of canonical transformations in classical mechanics to unitary transformations of quantum mechanics (the latter being an analogue of field theory, its potential(s) being wave function whose square is then interpreted as probability density). For an already existing classical field theory, a similar recasting of (field theoretical) canonical transformations occurs due to the *second* quantization.

In conclusion we consider another method of definition of the classical Poisson brackets, which could be convenient in performance of the second quantization of physical fields. This method was proposed by Peierls (1952); we shall speak not about its original formulation, but about its generalization to the just discussed four-dimensionally symmetric approach.

Starting with the integral four-Lagrangian (7.2.37) and the action integral (7.2.39), let us consider the perturbed Lagrangian density

$$\mathfrak{L}' = \mathfrak{L} + \lambda \Phi \delta^{(4)}(x, x'). \tag{7.2.42}$$

Then the corresponding perturbation of the action integral is

$$S' = S + \lambda \Phi \tag{7.2.43}$$

where Φ is some *function* of canonical coordinates and momenta, λ being an infinitesimal parameter, and we shall consider only terms up to the first order of magnitude. Corresponding to the perturbation of the action, the field equations and their solutions should be modified too, and all these perturbations can be expanded in powers of λ. Then the first-order expression is

$$A'_a(x) = A_a(x) + \lambda \vec{D}_\Phi A_a(x). \tag{7.2.44}$$

The form of the perturbed Lagrangian (7.2.42) shows that the perturbation acts at the moment $t = t'$, thus if retarded solutions are used, the physical system should not 'know' about the perturbation before this moment (the old solutions are applicable). Relativistically speaking, the function $\vec{D}_\Phi A_a(x)$ should vanish outside the future light cone with the vertex at x'^μ. Similarly, it is also possible to consider advanced solutions:

$$A'_a(x) = A_a(x) + \lambda \overleftarrow{D}_\Phi A_a(x) \tag{7.2.45}$$

in which the function $\overleftarrow{D}_\Phi A_a(x)$ vanishes outside the past light cone with the vertex at x'^μ. One could as well introduce 'fuzzy' (in time) perturbations taking as Φ in (7.2.43) not a function, but integral[9] which however should not include regions infinitely remote in time (both in the past and in the future). Then instead of the light cone one should consider a hypersurface asymptotically tending to it, and instead of the exact vanishing of a function, the corresponding limiting process.

If one now considers instead of A_a some function of field variables Ψ, its perturbation can be written in a similar form

$$\Psi' = \Psi + \lambda \vec{D}_\Phi \Psi \tag{7.2.46}$$

in the first case, and

$$\Psi' = \Psi + \lambda \overleftarrow{D}_\Phi \Psi \tag{7.2.47}$$

in the second one. Peierls proposed to define Poisson's brackets using these perturbations of functions as

$$\{\Phi, \Psi\} = \vec{D}_\Phi \Psi - \overleftarrow{D}_\Phi \Psi. \tag{7.2.48}$$

Then the change of the integral four-Hamiltonian induced by a perturbation, is equal to

$$\Delta P_\beta = -\lambda \int \Phi \delta^{(4)}(x, x') dS_\beta. \tag{7.2.49}$$

Splitting the four-dimensional δ-function into covariant (multi-component) three-dimensional and (time-like) one-dimensional ones satisfying the relation

$$\delta^{(4)}(x, x') = \delta^\alpha(x, x')\delta_\alpha(t, t'), \tag{7.2.50}$$

we find

$$\Delta P_\beta = -\lambda \Phi \delta_\beta(t, t'). \tag{7.2.51}$$

---

[9]We emphasize that a function — not a functional — of canonical coordinates and momenta, was used according to the approach proposed by Peierls, even in the expression for the action integral.

Here the one-dimensional δ-function possesses the property

$$\int f(x)\delta_\beta(t, t')dx^\beta = f(x)|_{t=t'} \tag{7.2.52}$$

where t means not time, but a parameter which enumerates a family of hypersurfaces; thus the equality $t = t'$ means that the function (here, f) is taken on this hypersurface. When the Hamiltonian equation (7.2.5) is considered, we have for derivatives of the perturbed canonical momentum an expression

$$\mathfrak{P}'^{a\alpha}{}_{,\beta} = -\frac{\Delta^\alpha P'_\beta}{\Delta A_a} = \mathfrak{P}^{a\alpha}{}_{,\beta} + \lambda\frac{\Delta^\alpha \Phi}{\Delta A_a}\delta_\beta(t, t'). \tag{7.2.53}$$

This gives us derivatives of the perturbation of momentum which are proportional to the one-dimensional δ-function. This expression describes, as it was observed by Peierls, a sudden change of the canonical momentum

$$\Delta\mathfrak{P}^{a\alpha} = \lambda\frac{\Delta^\alpha \Phi}{\Delta A_a} \tag{7.2.54}$$

at the moment when the perturbation is switched on. Similarly, using the relation

$$\frac{\Delta^\alpha(\Delta P_\beta)}{\Delta \mathfrak{P}^{a\gamma}} = -\lambda\frac{\Delta^\alpha \Phi}{\Delta \mathfrak{P}^{a\gamma}}\delta_\beta(t, t') = \Delta A_{a,\beta}\delta^\alpha_\gamma, \tag{7.2.55}$$

following from the Hamiltonian equation (7.2.4), we find the value of a sudden change of the canonical coordinates when the perturbation is switched on:

$$\Delta A_a \delta^\alpha_\gamma = -\lambda\frac{\Delta^\alpha \Phi}{\Delta \mathfrak{P}^{a\gamma}}. \tag{7.2.56}$$

From the definition of variational derivative (7.3.7) it is obvious that then the change of a function of canonical momenta and coordinates Ψ when the perturbation is switched on, should take the form

$$\lambda\vec{D}_\Phi \Psi = \int \left[\frac{\Delta^\sigma \Psi}{\Delta A_a}\Delta_\Phi A_a + \frac{\Delta^\sigma \Psi}{\Delta \mathfrak{P}^{a\tau}}\Delta_\Phi \mathfrak{P}^{a\tau}\right] dS_\sigma. \tag{7.2.57}$$

Using (7.2.54) and (7.2.56), we recast this expression to

$$\vec{D}_\Phi \Psi = \int \left[\frac{\Delta^\tau \Psi}{\Delta A_a}\frac{\Delta_\Phi A_a}{\lambda}\delta^\sigma_t au + \frac{\Delta^\sigma \Psi}{\Delta \mathfrak{P}^{a\tau}}\frac{\Delta_\Phi \mathfrak{P}^{a\tau}}{\lambda}\right] dS_\sigma$$

$$= \int \left[\frac{\Delta^\tau \Phi}{\Delta A_a}\frac{\Delta^\sigma \Psi}{\Delta \mathfrak{P}^{a\tau}} - \frac{\Delta^\sigma \Phi}{\Delta \mathfrak{P}^{a\tau}}\frac{\Delta^\tau \Psi}{\Delta A_a}\right] dS_\sigma = \{\Phi, \Psi\}. \tag{7.2.58}$$

As it was already observed above, physical quantities suffer a sudden change by switching on a perturbation; hence we have to admit that in the expression (7.2.58) the functions

Φ and Ψ are taken at the moments t_1 and $t_1 + 0$ respectively. If the same functions were taken at the moments t_1 and $t_1 - 0$, the perturbation had to disappear:

$$\overleftarrow{D}_\Phi \Psi = 0, \qquad (7.2.59)$$

as one should expect from the law of causality (the retardation of interactions). Perfectly analogous relations, which differ from the just obtained in a sign and the sequence of moments of time, hold when the advanced interaction is considered, thus, *e.g.*, for $\overleftarrow{D}_\Phi \Psi$ in the left-hand side of (7.2.58). Taking difference of the corresponding quantities, we obtain a universal expression for all moments of time and coinciding with Poisson's brackets (7.2.48), *quod erat demonstrandum*.

In the case of degenerate fields (*cf.* footnote 7.2), the field intensity $F_{b\alpha}$ has a number of independent non-trivial components which is *less* than that of the usual canonical velocity $A_{a,\alpha}$, because of symmetry properties of the quantities $F_{b\alpha}$ of which the Lagrangian density actually depends. Here it is impossible to resolve the equations

$$F_{a\alpha} = k^{b\beta}_{a\alpha} A_{b,\beta} \qquad (7.2.60)$$

with respect to $A_{a,\alpha}$. Then, introducing alongside with A_a and $\mathfrak{P}^{a\alpha}$ new quantities $\Xi^\alpha_{b\beta}$ and $\Sigma^{b\beta}$,

$$\begin{aligned}
\Xi^\alpha_{b\beta} &:= k^{a\alpha}_{b\beta} A_a; \\
\Sigma^{b\beta} &:= \frac{\partial \mathfrak{L}}{\partial F_{b\beta}};
\end{aligned} \right\} \qquad (7.2.61)$$

$$\mathfrak{P}^{a\alpha} := k^{a\alpha}_{b\beta} \Sigma^{b\beta}, \qquad (7.2.62)$$

we obtain instead of (7.2.51)

$$\delta P_\beta = \int \left[\Xi^\alpha_{b\mu,\beta} \delta \Sigma^{b\mu} - \mathfrak{P}^{a\alpha}{}_{,\beta} \delta A_a \right] dS_\alpha, \qquad (7.2.63)$$

so that the Hamiltonian equations in field theory take the form

$$\left. \begin{aligned}
\frac{\Delta^\alpha P_\beta}{\Delta \Sigma^{b\mu}} &= \Xi^\alpha_{b\mu,\beta}; \\
\frac{\Delta^\alpha P_\beta}{\Delta A_a} &= -\mathfrak{P}^{a\alpha}{}_{,\beta}
\end{aligned} \right\}. \qquad (7.2.64)$$

The second of these equations coincides (up to the number of non-trivial independent components of $\mathfrak{P}^{a\alpha}$) with (7.2.5), while the first one differs favorably from (7.2.4) since its left- and right-hand sides always have the same symmetry properties in the group of indices $_{b\mu}$. The further analysis of Poisson's brackets in the degenerate case is however rather cumbersome, and it requires individual approach to concrete fields. We shall not speak more about this case below.

The relations obtained in this section, are important in formulation of quantum field theory (in particular, in the next section where we however also consider another, independent approach to second quantization which is directly applicable to the case of degenerate fields).

7.3 Canonical Formalism and Quantization

The most direct way to implement a transition from classical to quantum theory is to apply the canonical quantization method. When it is applied to a system of massive points, one comes to the quantum (wave) mechanics: this quantization may be called the "primary" one. In such a transition the theory becomes based on the concept of probability waves instead of pointlike particles, thus one has now a field (in a certain sense, a continuos medium) distributed in space and changing with time. But when the canonical method is applied to (classical) physical fields or mechanical continuous media, one arrives, on the contrary, to the concept of particles, the excitation quanta of these systems. These are, in particular, also such "particles" as phonons existing exclusively on the background of a real mechanical continuum, while photons, fermions and other "true" particles exist on the background of vacua of the corresponding fields. This last type of quantization is called the *second quantization*.

Usually the canonical method merely reduces to an application of the quantum theoretical Poisson brackets, *i.e.*, a replacement of the classical Poisson brackets in relations written within a framework of the Hamiltonian (canonical) formalism of classical mechanics, to commutators with certain coefficients. Such a transition seems to be inevitable if we postulate that the key quantities of our theory are not c-numbers, but operators (q-numbers), so that they in general do not mutually commute. The scheme of such a transition is rather simple [see *e.g.* Dirac (1982); Leech (1958)] and it consists of the following.

The distributivity property of classical Poisson's brackets in mechanics (and field theory as well) reads

$$\{F, GH\} = G\{F, H\} + \{F, G\}H;$$

an equivalent formula reads

$$\{FH, G\} = F\{H, G\} + \{F, G\}H.$$

Here F, G and H are some functions (or functionals) of A_a and $\mathfrak{P}^{a\alpha}$. Let us now consider Poisson's brackets involving four such functions. Then we have, on one hand,

$$\{\Phi\Theta, \Psi\Xi\} = \Phi\{\Theta, \Psi\Xi\} + \{\Phi, \Psi\Xi\}\Theta =$$

$$\Phi\{\Theta, \Psi\}\Xi + \Phi\Psi\{\Theta, \Xi\} + \Psi\{\Phi, \Xi\}\Theta + \{\Phi, \Psi\}\Xi\Theta, \tag{7.3.1}$$

while on the other hand,

$$\{\Phi\Theta, \Psi\Xi\} = \{\Phi\Theta, \Psi\}\Xi + \Psi\{\Phi\Theta, \Xi\} =$$

$$\Phi\{\Theta, \Psi\}\Xi + \{\Phi, \Psi\}\Theta\Xi + \Psi\Phi\{\Theta, \Xi\} + \Psi\{\Phi, \Xi\}\Theta. \tag{7.3.2}$$

Equalizing these two expressions and performing simple identical transformations, we arrive to

$$[\Phi, \Psi]_-\{\Theta, \Xi\} = \{\Phi, \Psi\}[\Theta, \Xi]_- \tag{7.3.3}$$

where the notation
$$[\Phi, \Psi]_- := \Phi\Psi - \Psi\Phi \qquad (7.3.4)$$
is introduced for a commutator.

If we still are working in the classical theory, the commutators should naturally vanish, so that the relation (7.3.3) is satisfied trivially. However in the quantum theory, when the quantities appearing in a commutator, are operators, the necessary and sufficient condition for (7.3.3) to be satisfied, is a proportionality
$$\{\Phi, \Psi\} = \alpha[\Phi, \Psi]_-, \qquad (7.3.5)$$
such a relation having to hold for *any* pair of operators for which the classical Poisson brackets were determined in the classical theory. The proportionality coefficient α, in its turn, should be a *universal constant*, the same for any pair of operators.

Now it is easy to determine the dimensionality of this constant — it is
$$[\alpha] = [\mathcal{L}]^{-1}[l]^4 \qquad (7.3.6)$$
where $\mathcal{L}$ is a Lagrangian density and l, a distance. Since we have used here a natural system of units in which the velocity of light $c = 1$, we should in general revise this conclusion when working, *e.g.*, in the CGS system (for example, changing one of the four distance-like factors to time). Thus we may infer that
$$[\alpha] = (erg \cdot sec)^{-1}, \qquad (7.3.7)$$
the dimensionality of inverse *action*, very natural for the quantum theory. Next we shall see that α has to be purely imaginary. Let us suppose first that the classical Poisson brackets change under the complex conjugation (if we do not consider classical *matrix* quantities indeed) into
$$\{\Phi, \Psi\}^+ = \{\Phi^+, \Psi^+\}, \qquad (7.3.8)$$
the cross meaning, after the transition to quantum theory, the Hermitean conjugation. Then
$$\{\Phi, \Psi\}^+ = \alpha^*([\Phi, \Psi]_-)^+ = -\alpha^*[\Phi^+, \Psi^+]_- = -\frac{\alpha^*}{\alpha}\{\Phi^+, \Psi^+\}, \qquad (7.3.9)$$
so that a comparison with (7.3.8) yields
$$\alpha^* = -\alpha, \qquad (7.3.10)$$
i.e.
$$\alpha = -\frac{i}{\hbar}. \qquad (7.3.11)$$

The sign is chosen here for convenience (in the future applications), while the new real universal constant $\hbar$ has dimensionality of action ($erg \cdot sec$) and is interpreted as Planck's constant.

Since the new universal constant thus introduced in the theory, has a non-trivial dimensionality, it is worth "calibrating" our system of units in such a way that both velocity of light c and Planck's constant $\hbar$ would become equal to unity. We shall continue to call this system a natural one. Then the *quantum Poisson's brackets* take a simple form

$$\{\Phi, \Psi\} = -i[\Phi, \Psi]_-. \tag{7.3.12}$$

We see that this way of transition leads to commutation relations only, but not to anticommutation relations which could seem to be appropriate in the case of fermion fields.[10]

This is why Peierls (1952) proposed to use, if necessary, the operator of rotation by the complete angle 2π, assuming that such a rotation changes the sign of corresponding variables (field potentials, *i.e.* canonical coordinates). He proposed to admit that Poisson's brackets are then defined only for corresponding quantities multilplied by this operator — let us denote it as Θ (in general such a restriction of applicability of Poisson's brackets is not foreign to the spirit of our theory). Consider now Poisson's brackets for $\Theta\Phi$ and Ψ:

$$\{\Theta\Phi, \Psi\} = -i[\Theta\Phi, \Psi]_- = -i(\Theta\Phi\Psi - \Psi\Theta\Phi) \equiv$$
$$-i(\Theta\Phi\Psi - \Psi\Theta\Phi + \Phi\Theta\Psi - \Phi\Theta\Psi + \Phi\Psi\Theta - \Phi\Psi\Theta + \Psi\Phi\Theta - \Psi\Phi\Theta) =$$
$$-i([\Theta, \Phi]_+\Psi + [\Phi, \Psi]_+\Theta - \Psi[\Theta, \Phi]_+ - \Phi[\Theta, \Psi]_+). \tag{7.3.13}$$

If the operator Θ now anticommutes with both Φ and Ψ, while the anticummutator $[\Phi, \Psi]_+$ is supposed to be a c-number, and if one might consider in the left-hand side of (7.3.13) Θ merely as a factor from the left, a "division" by Θ yields

$$\{\Phi, \Psi\} = -i[\Phi, \Psi]_+. \tag{7.3.14}$$

However, one has to notice that Poisson's brackets in the left-hand side of (7.3.14) now cannot have any classical meaning, at least because they fail to possess the standard algebraic properties. The last circumstance should compel one to suppose that such a modification of "classical" mechanics (which could not be already called classical without quotation marks) is possible, which would include also fermionic particles with the "classical" variant of their statistics (as we have a classical limit for the bosonic particles). If such a new "classical" mechanics could be constructed, it should be, probably, called the corpuscular one, in a contrast to the wave (quantum) mechanics.

The above analysis is quite sufficient for a transition to obtaining concrete conclusions from quantum field theory. It is however helpful first to approach to the second quantization from a different viewpoint, taking into account some specific features of physical systems, as well as of coordinate transformations. This enables to relate this problem with the Noether theorem and simultaneously with the classical canonical formalism in field theory avoiding rather formal considerations like those used in derivation of the relation (7.3.5). This alternative way was briefly discussed in the monograph by Bogoliubov and Shirkov (1959), whose notations will be essentially used in our presentation of the subject.[11] At the

---

[10] It is important that such a suggestion can be related not to all constructions built of such fields. For example, if the relations involve dynamical variables (energy-momentum, spin, *etc.*), they should be namely the *commutation* relations.

[11] See also Schweber (1961) and Wentzel (1949).

same time, we shall try to investigate this way from somewhat different positions [cf. our previous publications: Mitskievich (1958c, 1959d)].

From the very beginning it is necessary to postulate that physical systems have to be described (in quantum theory) with the use of the amplitude of state, $\Phi[\Sigma]$, naturally determined on a hypersurface Σ fixing (relative) simultaneity of the world points on it. Let us suppose that the canonical coordinates (field potentials) and momenta are operators having an effect on the amplitude of state.

A mere transition to a new system of coordinates without changing the space-like hypersurface of simultaneity, cannot influence the amplitude of state, since a physical situation in the same reference frame (though described in alternative systems of coordinates) does not depend on the choice of this description, in particular, on how we choose to numerate space-time points. However a transition to another reference frame corresponds not only to a transformation of coordinates (if any; cf. the monad formalism), but also (essentially) to a new choice (one may say: calibration) of our hypersurface.[12] The amplitude of state will then in general suffer a change which can be expressed as

$$\Phi'[\Sigma'] = U[\Sigma, \Sigma']\Phi[\Sigma]. \qquad (7.3.15)$$

This transformation corresponds to a transition to a new system of coordinates *and* (simultaneously) to a new space-time section (new set of points) of which the new function(al)s will depend, while the coordinates of these new points in the new system of coordinates will numerically coincide with the coordinates of old points in the old system.[13] In other words, we now encounter the well known Lie derivative, $\pounds$ (we do not mention here the tangent vector to the reference frame congruence, like in $\pounds_\varsigma$, which is rather natural):

$$A_a \to A_a - \pounds A_a. \qquad (7.3.16)$$

Considering the corresponding transformation as an infinitesimal one, one may write

$$U[\Sigma] = I + \delta U[\Sigma] \qquad (7.3.17)$$

(in such a case of infinitesimality, it is natural to omit the second argument, Σ'); then

$$\Phi^+ A_a \Phi \to \Phi^+ A_a + \Phi^+ \left(\delta U^+ A_a + A_a \delta U - \pounds A_a\right) \Phi. \qquad (7.3.18)$$

If we admit that the matrix element $\Phi^+ A_a \Phi$ does not depend on the choice of the hypersurface which fixes simultaneity of events, we have to suppose that[14]

$$\pounds A_a = -\delta U A_a - A_a \delta U^+. \qquad (7.3.19)$$

---

[12]If a change of reference frame is described through the corresponding Lorentz transformations, as this is usual in special relativity. A more general reformulation using the monad formalism and curved space-time is straightforward only when normal monad congruences are considered.

[13]Here, a direct relation to the monad congruence seems to be transparent.

[14]Admitting that this takes place for arbitrary $\Phi[\Sigma]$.

Taking instead of A_a the identity matrix I, we shall not violate the relation (7.3.19), but the matrix δU proves then to be anti-Hermitian,

$$\delta U^+ = -\delta U; \qquad (7.3.20)$$

this corresponds to the property of the transformation matrix U to be unitary:[15]

$$U^+ U = I. \qquad (7.3.21)$$

Introducing for the convenience a Hermitian matrix δV,

$$\delta U = i\delta V, \ \ \delta V^+ = \delta V, \qquad (7.3.22)$$

we obtain from (7.3.19)

$$\pounds A_a = i[A_a, \delta V]_-. \qquad (7.3.23)$$

Since the operator δV is determined on a hypersurface, it is natural to take

$$\delta V = \int_\Sigma \delta \mathfrak{V}^\alpha dS_\alpha. \qquad (7.3.24)$$

Some of its properties can be found when the physical *vacuum* state described by the amplitude Φ_{vac}, is considered: this amplitude of state should not change from one hypersurface to another, due to conservation laws:

$$\Phi_{\text{vac}}[\Sigma_1] = \Phi_{\text{vac}}[\Sigma_2]. \qquad (7.3.25)$$

In another reference frame,

$$\Phi'_{\text{vac}}[\Sigma'_1] = \Phi'_{\text{vac}}[\Sigma'_2]. \qquad (7.3.26)$$

Thus

$$\delta V[\Sigma_1]\Phi_{\text{vac}}[\Sigma_1] = \delta V[\Sigma_2]\Phi_{\text{vac}}[\Sigma_2]. \qquad (7.3.27)$$

so that, using (7.3.25), we conclude that

$$\int_{\Sigma_1} \delta \mathfrak{V}^\alpha dS_\alpha = \int_{\Sigma_2} \delta \mathfrak{V}^\alpha dS_\alpha. \qquad (7.3.28)$$

Inverting direction of one of the normals and passing to a hypercylinder with infinitely distant lateral hypersurfaces, we obtain from the Gauss theorem

$$\oint_\Sigma \delta \mathfrak{V}^\alpha dS_\alpha = \int (\delta \mathfrak{V}^\alpha)_{,\alpha}(dx) = 0. \qquad (7.3.29)$$

---

[15] Unitary transformations in quantum theory correspond to the canonical transformations of classical mechanics (and, obviously, the field theory). Therefore they naturally have to express, in particular, the coordinate changes. See in this connection a paper by V. Fock given as an addendum to the Russian translation of the book by Dirac (1960).

Canonical Formalism

A natural generalization of this equality represents a differential conservation law

$$(\delta \mathfrak{V}^\alpha)_{,\alpha} = 0. \tag{7.3.30}$$

Our considerations here are not rigorous, but only heuristic ones. Therefore it is convenient to make use of the quasiclassical approximation, for which one may write

$$\Phi_1 = e^{iS} \Phi_0, \tag{7.3.31}$$

where the action integral S is taken in the "layer" between two space-like hypersurfaces:

$$S = \int_{\Sigma_1}^{\Sigma_2} \mathfrak{L}(dx). \tag{7.3.32}$$

Then

$$\Phi'_1 = e^{iS'} \Phi'_0. \tag{7.3.33}$$

Taking into account the transformation law,

$$\Phi'_1 = U[\Sigma_1]\Phi_1, \quad \Phi'_0 = U[\Sigma_0]\Phi_0, \tag{7.3.34}$$

it is easy to trace a chain of equalities

$$\Phi_1 = e^{iS}\Phi_0 = e^{iS}U[\Sigma_0]^{-1}\Phi'_0 = = e^{iS}U[\Sigma_0]^{-1}e^{-iS'}\Phi'_1 =$$
$$e^{iS}U[\Sigma_0]^{-1}e^{-iS'}U[\Sigma_1]\Phi'_1 \tag{7.3.35}$$

which — due to arbitrariness of the choice of Φ_1 — yield

$$e^{iS}U[\Sigma_0]^{-1}e^{-iS'}U[\Sigma_1] = I. \tag{7.3.36}$$

This relation can be conveniently put into form

$$e^{-iS'_1}U[\Sigma_1]e^{iS_1} = e^{-iS'_0}U[\Sigma_0]e^{iS_0} = \text{const}, \tag{7.3.37}$$

if the action integral (7.3.32) is expressed as $S = S_1 - S_0$, where

$$S_k = \int_{-\infty}^{\Sigma_k} \mathfrak{L}(dx). \tag{7.3.38}$$

Restricting ourselves to infinitesimal transformations, we have

$$S' = S + \delta S, \tag{7.3.39}$$

where S' differs from S primarily in the choice of Σ to Σ' (thus δ is essentially the Lie derivative); it is worth remembering the relation (7.3.17) too. Then, after rearranging (7.3.39) as

$$e^{-iS'} \approx e^{-iS}(1 - i\delta S), \tag{7.3.40}$$

the constant expression appearing in (7.3.37), takes the form

$$I + e^{-iS}(\delta U - i\delta S)e^{iS} = \text{const}. \tag{7.3.41}$$

It should be identified with the unit matrix, hence

$$\delta U = i\delta S. \tag{7.3.42}$$

Thus, comparing (7.3.42) with (7.3.22) and taking into account the conservation law (7.3.30), we have to identify δV and δS, the first quantity being an integral over a closed hypersurface, while the second one is the action integral (over a four-dimensional volume surrounded by this hypersurface). Restricting ourselves to a purely heuristic level of considerations, we observe that the construction satisfying a weak (hence, full-fledged in the sense of physics) conservation law and besides being a differential concomitant of $\mathfrak{L}$, is well known: this is, up to infinitesimal factor ϵ, $\mathfrak{w}^\alpha$,

$$\delta \mathfrak{V}^\alpha = \epsilon \left(-\mathfrak{t}^\alpha_\sigma \xi^\sigma + \mathfrak{M}^{\alpha\tau}_\sigma \xi^\sigma{}_{,\tau}\right). \tag{7.3.43}$$

Then

$$\pounds_\xi A_a = -i \left[A_a, \int dS_\alpha \left(\mathfrak{t}^\alpha_\sigma \xi^\sigma - \mathfrak{M}^{\alpha\tau}_\sigma \xi^\sigma{}_{,\tau}\right)\right]_-. \tag{7.3.44}$$

In its turn the left-hand side of this expression can be rewritten as $\pounds_\xi A_a = A_{a,\sigma}\xi^\sigma - A_a|^\tau_\sigma \xi^\sigma{}_{,\tau}$; hence the relation (7.3.44) takes the form

$$\int dS^{(x')}_\alpha \left\{\left(A_{a,\sigma}\xi^\sigma - A_a|^\tau_\sigma \xi^\sigma{}_{,\tau}\right)\delta^\alpha(x,x')\right.$$
$$\left. + i\left[A_a(x), \left(\mathfrak{t}^\alpha_\sigma(x')\xi^\sigma(x') - \mathfrak{M}^{\alpha\tau}_\sigma(x')\xi^\sigma(x')_{,\tau}\right)\right]_-\right\} = 0. \tag{7.3.45}$$

Since the vector field ξ (the transformation of coordinates) is arbitrary, the relation (7.3.45) splits into two parts:

$$A_{a,\sigma}\delta^\alpha(x,x') = -i\left[A_a(x), \mathfrak{t}^\alpha_\sigma(x')\right]_- \tag{7.3.46}$$

and

$$A_a|^\tau_\sigma \delta^\alpha(x,x')n_\alpha = -i\left[A_a(x), n_\alpha \mathfrak{M}^{\alpha\tau}_\sigma(x')\right]_-. \tag{7.3.47}$$

Here in both cases all quantities are taken on the same hypersurface. If we perform integration over this hypersurface, we obtain in accordance with (7.2.14) and (7.2.22) (where (7.3.12) should be taken into account)

$$A_{a,\sigma} = -i[A_a, P_\sigma]_- \tag{7.3.48}$$

and

$$A_a|^\tau_\sigma = -i[A_a, S^\tau_\sigma]_-. \tag{7.3.49}$$

So we have come once more to the quantum theoretical Poisson brackets, though now from a completely different side: these brackets should be written not for arbitrary quantities, but necessarily only for (1) dynamical variables (energy-momentum four-'vector', spin 'tensor' or the corresponding densities) and (2) canonical variables (field potentials) or constructions thereof.

Thus potentials of fields, as well as their derivatives, now are operators acting on the amplitude of state. All fields including the metric one, are considered here on the same basis. However the metric field simultaneously serves for geometric needs — in particular, it determines space- or time-like, or null character of any vector. Therefore quantization of gravity immediately leads to a revision of the concept of space-like hypersurfaces, which is of such an importance in description of integral physical quantities (energy-momentum, spin) and in the very formulation of the second quantization procedure with the use of Poisson's brackets as well. One may propose different ways to overcome this difficulty, for example, via introduction of the second (non-physical, thus not subjected to quantization) metric tensor in the bimetric formalism.[16]

---

[16] All such ways and methods used by different authors, still did not resolve the problem of quantization in general relativity; see (Kiefer 2004).

Chapter 8

Concluding Remarks

In this book it was of course impossible to characterize the whole magnitude of fundamental problems and applications of the reference frames theory in general and special relativity, including their new development tendencies. For example, we have deliberately ignored here many details of the case of normal congruences (non-rotating reference frames) which are considered to be so important in building the canonical formalism in general relativity to the end of gravity quantization.[1] Elementary ideas of quantization are discussed in (Mitskievich 1969) and fundamentals of the canonical formalism, in (Misner, Thorne and Wheeler 1973, Mitskievich, Yefremov and Nesterov 1985, Kiefer 2004). It is however worth mentioning here some works in alternative directions which could be helpful in understanding the development tendencies in these areas.

Major difficulty in realization of canonical formalism in field theory is due to the problem of equations of constraints. With help of gauge theory ideas, and in particular those of the Yang–Mills theory, Ashtekar (1988a) succeeded in making exceptionally good choice of field canonical variables and applying them to different types of problems [*e.g.* (Ashtekar 1986, Ashtekar 1988b, Ashtekar 1989)]. One of trends in the development of this theory consisted in a study of spaces with self-dual Weyl tensor (Koshti and Dadhich 1990) for which a set of general theorems was obtained [in a more conventional approach, such spaces are often called 'heavenly spaces' ($\mathcal{H}$-spaces); they were studied and applied to solving Einstein's equation, especially by E. Newman, J. Plebański, and D. Finley]. The very foundations of the gravitational field theory were also considered, and among them the ac-

---

[1]In fact, the author is by no means sure that this straightforward way could lead to a real solution of the problem. The search should be aimed at some approach which does not rely on any classical background; instead it has to be genuinely quantum one — moreover, not only the whole classical theory should follow from it, but even the existing quantum theory should represent a kind of its approximation. Its elementary concepts should be rather topological than geometrical (the space-time geometry has to be a secondary product and not its immediate base). And even more: we have to get away from our present-day dependence on phenomenological parameters (such as 'fundamental' constants adjusted to any kind of experimental data); instead of them some abstract axioms should be introduced directly yielding a new theory which would unify all interactions in physics in their evolution, in microworld as well as in the cosmological sense. The first steps in this direction can be found in (Efremov and Mitskievich 2003a, 2003b), (Efremov, Mitskievich and Hernández Magdaleno 2004), and (Efremov, Mitskievich, Hernández Magdaleno and Serrano Bautista 2005).

tion principle with the Palatini approach (Capovilla, Dell, Jacobson and Mason 1991, Floreanini and Percacci 1990) and the covariant action formulation (Jacobson and Smolin 1988). Sometimes new fruitful developments were achieved in the theory of initial data with the use of conformally flat spaces and space-time complexification (Wagh and Saraykar 1989).

The attention was focused in this direction mainly on the canonical formalism in gravitation theory where vacuum constraints were brought to a solvable formulation (Robinson and Soteriou 1990) and canonical transformations were introduced together with the corresponding generators in the field theory (Dolan 1989). Ashtekar's formalism was connected with the tetrad approach (Henneaux, Nelson and Schomblond 1989), then a first order tetrad form was introduced for description of the action principle and formulation of the canonical theory (Kamimura and Fukayama 1990), and some elegant results were obtained in the further use of three-spinors in Ashtekar's formalism (Perjés 1990).

Several papers, including the initial works of Ashtekar himself, were dedicated to development of methods of 3+1-splitting of space-time (Ashtekar 1988b, Ashtekar, Jacobson and Smolin 1988, Wallner 1990). The energy problems (*e.g.*, the localizability of energy in general relativity) were also considered on the basis of canonical formalism (Nester 1991).

Another branch of these studies was directed to obtaining new solutions of the gravitational field equations with help of specific choices of Ashtekar's canonical variables. Hereby solutions for gravitational instantons were found (Samuel 1988), spherically symmetric cosmological models (Koshti and Dadhich 1989) as well as a wide class of spherically symmetric problems of general relativity were studied (Bengtsson 1990), and the latter paper offers a general discussion of further prospects in applications of Ashtekar's formalism which is of considerable independent interest. The formalism was applied to the Kantowski-Sachs type solutions with a perfect fluid (Bombelli and Torrence 1990) as well as to cosmological models of different Bianchi types (Kodama 1988) (here quantization of these models was also with the use of Ashtekar's canonical variables). It is significant that Ashtakar's formalism in general does not imply non-degeneracy of the metric tensor, so that perfectly new solutions, inadmissible from the standpoint of the traditional theory, enter the scope of the study. For example, several solutions with massive scalar field were thus obtained without explicit use of the metric (Peldán 1990).

As we see, even a single innovation in the formulation of the canonical theory in general relativity (which can be related to applications of reference frames) showed an explosion-like evolution in few years and provoked a great number of fine results. This is not always the reference frame trend proper, of the type which was discussed in our book. The authors go often beyond the limits of physical reference frames while considering either complexified spaces (such as self-dual ones which may have both coordinated, tetrad or spinor representations), or null congruences. But in all the cases, far-reaching analogies clearly emerged between the theories of gravitation and electromagnetism (or Yang-Mills theory), and they were often used for determination of new directions of research and formulation of new fruitful approaches. It is worth repeating that we tried to give in this book a review of quite a general part of the reference frames formalism from the viewpoint of description of observables and interpretation of physical effects which are calculated initially, as a rule, in a four-dimensional form related only to the intrinsic symmetry properties of concrete problems without connection to specific reference frames, but the latter ones proved to be essential in the final physical interpretation of the results.

Returning to the approach to the reference frames formalism discussed in this book, I hope to have clearly shown that it differs from the previous approaches in a more consistent use of the powerful techniques of differential geometry. At the same time, new and more convenient notations were introduced which are as near to the conventional ones of the simple three-dimensional vector calculus as it was possible to achieve. The reader might even take our notations for those conventional ones and consider them to be not general covariant. As a matter of fact, they *are* covariant under arbitrary transformations of coordinates, and the distinction between different reference frames is to be exclusively seen in the use of a concrete monad τ which is of course a real four-(co)vector. The advantage of such notations (as div or curl) is in greatly facilitated interpretation possibilities when one considers equations and effects of relativistic physics in arbitrary reference frames.

In particular, I hope to have shown that gravitoelectromagnetism by no means is a hypothesis but a strict consequence of Einstein's general relativity. In fact, it even is a paraphrase for a significant part of the standard gravitation theory. Consequently, our task is not so much to verify the theory from the experimental viewpoint, but to refine the experimental means in physics up to this new level. We are studying fundamental problems of general relativity to the end of better understanding of this theory; its most exotic features clearly and vividly show its profound implications, its boundaries, and critical areas of growth of our knowledge.

References

Anderson, J.L. (1958a) *Phys. Rev.* **110**, 1197.

Anderson, J.L. (1958b) *Phys. Rev.* **111**, 965.

Anderson, J.L. (1959) *Phys. Rev.* **114**, 1182.

Ashby, N., and Shahid-Saless, B. (1990) *Phys. Rev. D* **42**, 1118.

Ashtekar, A. (1986) *Phys. Rev. Lett.* **57**, 2244.

Ashtekar, A. (1988a) *New Perspectives in Canonical Gravity* (Napoli: Bibliopolis).

Ashtekar, A. (1988b) *Contemporary Mathematics* **71**, 39.

Ashtekar, A. (1989) In: *Proc. 9th Int. Congr. Math. Phys.,* Swansea, 17-27 July, 1988 (Bristol, New York: AMS), p. 286.

Ashtekar, A., Jacobson, T., and Smolin, L. (1988) *Commun. Math. Phys.* **115**, 631.

Bauer, H. (1918) *Zs. Physik* **19**, 163.

Bel, Ll. (1959) *Compt. Rend. Paris* **248**, 2161.

Bengtsson, I. (1990) *Class. Quantum Grav.* **7**, 27.

Bogoliubov, N.N., and Shirkov, D.V. (1959) *Introduction to the Theory of Quantized Fields* (New York, London: Interscience Publishers).

Bohr, N., and Rosenfeld, L. (1933) *Kgl. Danske Videnskab. Selskab., Mat.-fys. Medd.* **12**, 8.

Bombelli, L., and Torrence, R.J. (1990) *Class. Quantum Grav.* **7**, 1747.

Bonnor, W.B. (1969) *Commun. Math. Phys.* **13**, 163.

Bowler, M.G. (1976) *Gravitation and Relativity* (Oxford: Pergamon).

Boyer, T.H. (1980) In: *Foundations of Radiation Theory and Quantum Electrodynamics* (New York, London: Plenum), p. 49.

Brezhnev, V.S. (1964) *Izv. Vuzov, Ser. Phys.* No. **6**, 77 (in Russian).

Brill, D. (1972) In: *Methods of Local and Global Differential Geometry in General Relativity* (Berlin: Springer), p. 45.

Brillouin, L. (1970) *Relativity Reexamined* (New York: Academic Press).

Capovilla, R., Dell, J., Jacobson, T., and Mason, L. (1991) *Class. Quantum Grav.* **8**, 41.

Carmeli, M. (2000) *Group Theory and General Relativity* (World Scientific, Singapore).

Cattaneo, C. (1958) *Nuovo Cim.* **10**, 318.

Cattaneo, C. (1959) *Ann. mat. pura e appl.* **48**, 361.

Cattaneo, C. (1961) *Rend. mat. e appl.* **20**, 18.

Cattaneo, C. (1962) *Rend. mat. e appl.* **21**, 373.

Chandrasekhar, S. (1998) *The Mathematical Theory of Black Holes* (Oxford: Clarendon Press).

Choquet-Bruhat, Y., DeWitt-Morette, C., and Dillard-Bleick, M. (1982) *Analysis, Manifolds, and Physics* (Amsterdam: North-Holland).

Cotton, E. (1899) *Ann. Fac. Sci. Toulouse II* **1**, 385.

Debever, R. (1959) *Comptes Rendus Paris* **249**, 1744.

Dehnen, H. (1962) *Zs. Naturforsch.* **17a**, 18.

Dehnen, H., Hönl, H., and Westpfahl, K. (1961) *Zs. Physik* **164**, 483.

DeWitt, B.S. (1965) *Dynamical Theory of Groups and Fields* (New York: Gordon and Breach).

DeWitt, B.S. (1966) *Phys. Rev. Lett.* **16**, 1092.

DeWitt, B.S. (2003) *The Global Approach to Quantum Field Theory*, vol. 1 & 2. (Oxford: Clarendon Press).

Dirac, P.A.M (1947) *The Principles of Quantum Mechanics* (Oxford: Clarendon Press) [Russian translation (1960)].

Dirac, P.A.M (1964) *Lectures on Quantum Mechanics* (New York: Yeshiva University).

Dolan, B.P. (1989) *Phys. Lett. B* **233**, 89.

Eckart, C. (1940) *Phys. Rev.* **58**, 919.

Efremov, V.N., and Mitskievich, N.V. (2003a) *Discrete model of spacetime in terms of inverse spectra of the T_0 Alexandroff topological spaces*, e-print gr-qc/0301063.

Efremov, V.N., and Mitskievich, N.V. (2003b) *A T_0-discrete universe model with five low-energy fundamental interactions and the coupling constants hierarchy*, e-print gr-qc/0309133.

Efremov, V.N., Mitskievich, N.V., and Hernández Magdaleno, A.M. (2004) *Gravitation & Cosmology* **10**, 201.

Efremov, V.N., Mitskievich, N.V., Hernández Magdaleno, A.M., and Serrano Bautista, R. (2005) *Class. Quantum Grav.* **22**, 3725.

Eguchi, T., Gilkey, P.B., and Hanson, A.J. (1980) *Phys. Reports* **66**, 213.

Ehlers, J. (1961) *Akad. Wiss. Lit. Mainz, Abhandl. Math.-Nat. Kl.* Nr. **11**.

Eisenhart, L.P. (1926) *Riemannian Geometry* (Princeton, N.J.: Princeton University Press).

Eisenhart, L.P. (1933) *Continuous Groups of Transformations* (Princeton, N.J.: Princeton University Press).

Eisenhart, L.P. (1972) *Non-Riemannian Geometry* (Providence, RI: American Mathematical Society).

Epikhin, E.N., Pulido, I., and Mitskievich, N.V. (1972) In: *Abstracts of the reports presented at the 3rd Soviet Gravitational Conference* (Erevan: University Press), p. 380.

Floreanini, R., and Percacci, R. (1990) *Class. Quantum Grav.* **7**, 1805.

Fock, V.A. (1964) *The Theory of Space, Time, and Gravitation* (New York: MacMillan).

Fock, V.A. (1971) *Voprosy filosofii [Philosophy Problems]*, No. 3, 46. In Russian.

García, A.A., Hehl, F.W., Heinicke, Ch., and Macías, A. (2004) *Class. Quantum Grav.* **21**, 1099.

Géhéniau, J. (1957) *Comptes Rendus Paris* **244**, 723.

Gödel, K. (1949) *Revs. Mod. Phys.* **21**, 447.

Goldstein, H. (1980) *Classical Mechanics* (Reading, London: Addison–Wesley).

Goldstein, S. (1976) *Lectures in Fluid Mechanics* (Providence, R.I.: AMS).

Gorbatsievich, A.K. (1985) *Quantum Mechanics in General Relativity* (Minsk: Universitetskoie). In Russian.

Hawking, S.W., and Ellis, G.F.R. (1973) *The Large Scale Structure of Space-Time* (Cambridge: Cambridge University Press).

Henneaux, M., Nelson, J.E., and Schomblond, C. (1989) *Phys. Rev.* D**39**, 434.

Hönl, H., and Soergel-Fabricius, Chr. (1961) *Zs. Phys.* **163**, 571.

Horský, J., and Mitskievich, N.V. (1989). *Czech. J. Phys. B*, **39**, 957.

Hughston, L.P. (1979) *Twistors and Particles* (Berlin, Heidelberg, New York: Springer-Verlag).

Islam, J.N. (1985) *Rotating fields in general relativity* (Cambridge, Cambridge University Press).

Israel, W. (1970) *Commun. Dublin Inst. Adv. Studies* **A19**, 1.

Ivanitskaya, O.S. (1969) *Generalized Lorentz Transformations and Their Applications* (Minsk: Nauka i tekhnika). In Russian.

Ivanitskaya, O.S. (1979) *Lorentzian Basis and Gravitational Effects in Einstein's Gravitation Theory* (Minsk: Nauka i tekhnika) In Russian.

Ivanitskaya, O.S., Mitskievich, N.V., and Vladimirov, Yu.S. (1985) *Reference Frames in General Relativity.* Preprint No. 374, Inst. of Physics, Acad. Sci. BSSR, Minsk.

Ivanitskaya, O.S., Mitskievich, N.V., and Vladimirov, Yu.S. (1986) In: *Relativity in Celestial Mechanics and Astrometry* (Dordrecht: Reidel). Pp. 177-186.

Jacobson, T., and Smolin, L. (1988) *Class. Quantum Grav.* **5**, 583.

Jantzen, R.T., Carini, P., and Bini, D. (1992) *Ann. Phys. (USA)* **219**, 1.

Kamimura, K., and Fukayama, T. (1990) *Phys. Rev. D* **41**, 1885.

Khaikin, S.E. (1947) *Mechanics*, 2nd edition (Moscow, Leningrad: GITTL). In Russian.

Kiefer, C. (2004) *Quantum Gravity* (Oxford: Clarendon Press).

Knapecz, G. (1959) *Ann. Phys. (Leipzig)* **3**, 340.

Knapecz, G. (1960) *Ann. Phys. (Leipzig)* **6**, 44.

Kodama, H. (1988) *Prog. Theoret. Phys.* **80**, 1024.

Koshti, S., and Dadhich, N. (1989) *Class. Quantum Grav.* **6**, L223.

Komar, A.B. (1958) *Phys. Rev.* **111**, 1182.

Komar, A.B. (1959) *Phys. Rev.* **113**, 934.

Koshti, S., and Dadhich, N. (1990) *Class. Quantum Grav.* **7**, L5.

Kramer, D., Stephani, H., MacCallum M., and Herlt, E. (1980) *Exact Solutions of Einstein's Field Equations* (Berlin: Deutscher Verlag der Wissenschaften; Cambridge, UK: Cambridge University Press).

Lanczos, C. (1938) *Ann. Math.* **39**, 842.

Lanczos, C. (1962) *Revs. Mod. Phys.* **34**, 379.

Landau, L.D., and Lifshitz, E.M. (1976) *Theoretical Physics, Vol. I: Mechanics* (Oxford: Pergamon Press).

Landau, L.D., and Lifshitz, E.M. (1971) *Classical Field Theory* (Reading, Mass.: Addison-Wesley).

Landau, L.D., and Lifshitz, E.M. (1973) *Theoretical Physics, Vol. II: Field Theory* (Moscow: Nauka), in Russian.

Leech, J.W. (1958) *Classical mechanics* (New York: Wiley and Sons).

Levashyov, A.E. (1979) *Motion and Duality in Relativistic Electrodynamics* (Minsk: Byelorussian Univ. Press), in Russian.

Lichnerowicz, A. (1955) *Théories relativistes de la gravitation et de l'électromagnétisme* (Paris: Masson).

Lichnerowicz, A. (1960) *Ann. di Mat. Pura e Appl.* **50**, 1.

Lief, B. (1951) *Phys. Rev.* **84**, 345.

Lovelock, D. (1971) *J. Math. Phys.* **12**, 498.

Macdonald, D., and Thorne, K.S. (1982) *Monthly Not. Roy. Astr. Soc.* **198**, 345.

Marck, J.-A. (1983) *Proc. Roy. Soc. London A* **385**, 431.

Massa, E. (1974) *Gen. Relat. and Grav.* **5**, 555.

Massa, E. (1974) *Gen. Relat. and Grav.* **5**, 573.

Massa, E. (1974) *Gen. Relat. and Grav.* **5**, 715.

Matte, A. (1953) *Canad. J. Math.* **5**, 1.

Meier, W., and Salié, N. (1979) *Theoret. and Math. Phys.* **38**, 408. In Russian.

Misner, C.W., Thorne, K.S., and Wheeler, J.A. (1973) *Gravitation* (San Francisco: W.H. Freeman).

Mitskievich (Mizkjewitsch), N.V. (1958) *Ann. Phys. (Leipzig)* **1**, 319. In German.

Mitskievich, N.V. (1965) *Izvestiya Vuzov, Phys.* No. **6**, 87. In Russian.

Mitskievich, N.V. (1969) *Physical Fields in General Relativity* (Moscow: Nauka). In Russian.

Mitskievich, N.V. (1972) In: *Einsteinian Collection 1971* (Moscow: Nauka), p. 67. In Russian.

Mitskievich, N.V. (1975) In: *Problems of Gravitation Theory* (Erevan: University Press), p. 104. In Russian.

Mitskievich, N.V. (1976) In: *Problems of Theory of Gravitation and Elementary Particles* (Moscow: Atomizdat). Fasc. 7, p. 15. In Russian.

Mitskievich, N.V. (1979) *Supplementary Chapter* to the Russian Translation (Moscow: Mir) of the Book: Bowler (1976). In Russian.

Mitskievich, N.V. (1981a) *Experim. Technik d. Phys.* **29**, 213.

Mitskievich, N.V. (1981b) In: *Abstracts of the reports presented at the 5th Soviet Gravitational Conference* (Moscow: Moscow University Press), p. 43. In Russian.

Mitskievich, N.V. (1983) *Proc. Einstein Found. Internat.* **1**, 137.

Mitskievich, N.V. (1989) In: *Gravitation and Waves. Transactions of Inst. of Physics, Estonian Acad. Sci., Tartu* **65**, p. 104.

Mitskievich, N.V. (1990) In: *Differential Geometry and Its Applications* (Singapore: World Scientific), p. 297.

Mitskievich, N.V. (1991) Reference frames: description and interpretation of effects of the relativistic physics. In: *Classical Field Theory and Gravitation Theory* (The Progress in Science and Technology, Vol. 3), VINITI, Moscow, 1991, pp. 108-165. In Russian.

Mitskievich, N.V. (1996) *Relativistic Physics in Arbitrary Reference Frames.* Book e-print gr-qc/9606051, 137 pp.

Mitskievich, N.V. (1999a) *Int. J. Theor. Phys.* **38**, 997.

Mitskievich, N.V. (1999b) *Gen. Rel. Grav.* **31**, 713.

Mitskievich, N.V. (2001) *Lorentz Force Free Charged Fluids in General Relativity: The Physical Interpretation.* In: *Exact Solutions and Scalar Fields in Gravity: Recent Developments* (New York: Kluwer Academic/Plenum Publishers), p. 311.

Mitskievich, N.V. (2003) *Revista Mexicana de Física* **49** Supl. 2, 39.

Mitskievich, N.V., and Cindra, J.L. (1988) In: *Current problems of Quantum Mechanics and Statistical Physics* (Moscow: Peoples' Friendship University Press), p. 89. In Russian.

Mitskievich, N.V., and Gupta, S. (1980) In: *Abstracts of contributed papers, 9th International Conference of GRG* (Jena: Friedrich Schiller University Press), p. 190.

Mitskievich, N.V., and Kalev, D.A. (1975) *Comptes Rendus Acad. Bulg. Sci.* **28**, 735. In Russian.

Mitskievich, N.V., and Kumaradtya, K.K. (1989) *J. Math. Phys.* **30**, 1095.

Mitskievich, N.V., and López-Benítez, L.I. (2004) *Grav. & Cosmol.* **10**, 127.

Mitskievich, N.V., and Merkulov, S.A. (1985) *Tensor Calculus in Field Theory* (Moscow: Peoples' Friendship University Press). In Russian.

Mitskievich, N.V., and Mujica, J.D. (1968) *Doklady Akad. Nauk SSSR* **176**, 809. In Russian.

Mitskievich, N.V., and Nesterov, A.I. ((1981) *Exper. Techn. der Physik* **29**, 333.

Mitskievich, N.V., and Nesterov, A.I. (1991) *Class. Quantum Grav.* **8**, L45.

Mitskievich, N.V., and Pulido, I. (1970) *Doklady Akad. Nauk SSSR* **192**, 1263. In Russian.

Mitskievich, N.V., and Ribeiro Teodoro, M. (1969) *Sov. Phys. JETP* **29**, 515.

Mitskievich, N.V., and Tsalakou, G.A. (1991) *Class. Quantum Grav.* **8**, 209.

Mitskievich, N.V., and Uldin, A.V. (1983) *A gravitational lens in the Kerr field, Deponent No. 2654-83* (Moscow: Institute of Scientific and Technical Information, Academy of Sciences of USSR) In Russian.

Mitskievich, N.V., Yefremov, A.P., and Nesterov, A.I. (1985). *Dynamics of Fields in General Relativity* (Moscow: Energoatomizdat) In Russian.

Mitskievich, N.V., and Zaharow, V.N. (1970) *Doklady Akad. Nauk SSSR* **195**, 321. In Russian.

Nester, J.M. (1991) *Class. Quantum Grav.* **8**, L19.

Newton, Sir Isaac (1962) *Mathematical Principles of Natural Philosophy.* (Ed. and comm. by F. Cajori.) Vol. 1 & 2. (Berkeley and Los Angeles, University of California Press).

Noether, E. (1918) *Götting. Nachr.*, **235**.

Nordtvedt, K. (1988) *Int. J. Theoret. Phys.* **27**, 1395.

Palatini, A. (1919) *Rend. circolo mat. Palermo* **43**, 203.

Papini, G. (1966) *Physics Letters* **23**, 418.

Papini, G. (1969) *Nuovo Cim.* **63 B**, 549.

Peebles, P.J.E. (1993) *Principles of Physical Cosmology* (Princeton: Princeton University Press).

Peebles, P.J.E. (1997) In: Bahcall, J.N., and Ostriker, J.P., Editors, *Unsolved Problems in Astrophysics* (Princeton: Princeton University Press), p. 4.

Peldán, P. (1990) *Phys. Lett. B* **248**, 62.

Penrose, R. (1960) *Ann. Phys. (USA)* **10**, 171.

Penrose, R., and Rindler, W. (1984a) *Spinors and Space-Time. Vol. 1. Two-Spinor Calculus and Relativistic Fields* (Cambridge: Cambridge University Press).

Penrose, R., and Rindler, W. (1984b) *Spinors and Space-Time. Vol. 2. Spinor and Twistor Methods in Space-Time Geometry.* (Cambridge: Cambridge University Press).

Peierls, R. (1952) *Proc. Roy. Soc A.* **214**, 143.

Peres, A. (1959) *Phys. Rev. Lett.* **3**, 571.

Perjés, Z. (1990) *The Parametric Manifold Picture of Space-Time.* Preprint, Central Research Inst. for Physics, Budapest, Hungary.

Petrov, A.Z. (1966) *New Methods in General Theory of Relativity* (Moscow: Nauka). In Russian.

Pirani, F.A.E. (1957) *Phys. Rev.* **105**, 1089.

Pirani, F.A.E, and Schild, A. (1961) *Bulletin de l'Académie Polonaise des Sciences, Série de sciences math., astr. et phys.* **IX**, No. 7, 543.

Regge, T. (1958) *Nuovo Cimento* **7**, 215.

Robinson, D.C., and Soteriou, C. (1990) *Class. Quantum Grav.* **7**, L247.

Rodichev, V.I. (1972) In: *Einsteinian Collection 1971* (Moscow: Nauka). In Russian.

Rodichev, V.I. (1974) *Theory of Gravitation in Orthonormal Tetrad Frames* (Moscow: Nauka). In Russian.

Ryan, M.P., and Shepley, L.C. (1975) *Homogeneous Relativistic Cosmologies* (Princeton, N.J.: Princeton University Press).

Rylov, Yu.A. (1964) *Ann. der Phys. (Leipzig)* **7**, 12.

Sachs, R.K., and Wu, H. (1977) *General Relativity for Mathematicians* (New York, Heidelberg, Berlin: Springer).

Sakina, K., and Chiba, J. (1980) *Lett. N. Cim.* **27**, 184.

Sakurai, J.J. (1960). *Ann. Phys. (New York)*, **11**, 1.

Salié, N. (1986) *Astron. Nachr.* **307**, 335.

Samuel, J. (1988) *Class. Quantum Grav.* **5**, L123.

Schmutzer, E. (1968) *Relativistische Physik (klassische Theorie)* (Leipzig: Teubner).

Schmutzer, E. (1975) *Scr. Fac. Sci. Nat. UJEP Brunensis, Physica* **3-4**, 5, 279.

Schmutzer, E. (1989) *Grundlagen der Theoretischen Physik* (Mannheim, Wien, Zürich: BI-Wissenschaftsverlag), Teile 1 & 2.

Schmutzer, E. (2004) *Projektive Einheitliche Feldtheorie mit Anwendungen in Kosmologie und Astrophysik* (Frankfurt am Main: Wissenschaftlicher Verlag Harry Deutsch GmbH). Mit einem Anhang von A.K. Gorbatsievich.

Schmutzer, E., and Plebański, J.F. (1977) *Fortschr. d. Physik* **25**, 37.

Schmutzer, E., und Schütz, W. (1983) *Galileo Galilei*, 5. Auflage (Leipzig: Teubner).

Schouten, I.A., and Struik, D.J.(1935) *Einführung in die neueren Methoden der Differentialgeometrie.* Vol. 1 (Gronongen: Noordhoff).

Schouten, I.A., and Struik, D.J. (1938) *Einführung in die neueren Methoden der Differentialgeometrie.* Vol. 2 (Gronongen: Noordhoff).

Schrödinger, E. (1956) *Expanding Universes* (Cambridge: Cambridge University Press).

Schutz, B.F. (1980) *Geometrical Methods of Mathematical Physics* (Cambridge: Cambridge University Press).

Schweber, S.S. (1961) *Introduction to Relativistic Quantum Field Theory* (Evanston, Ill.: Row and Peterson).

Shteingrad, Z.A. (1974) *Doklady Akad. Nauk SSSR* **217**, 1296. In Russian.

Stephani, H. (1990) *General Relativity*, 2nd edition (Cambridge, UK: Cambridge University Press).

Stephani, H., Kramer, D., MacCallum, M., Hoenselaers, C., and Herlt, E. (2003) *Exact Solutions of Einstein's Field Equations*, 2nd edition (Cambridge: Cambridge University Press).

Synge, J.L. (1960) *Relativity: The General Theory* (Amsterdam: North-Holland).

Synge, J.L. (1965) *Relativity: The Special Theory* (Amsterdam: North-Holland).

Synge, J.L. (1974) *Hermathena (a Dublin University review)*. No. cxvii (without pagination).

Taub, A.H. (1978) *Ann. Rev. Fluid Mech.* **10**, 301.

Thorne, K.S., and Macdonald, D. (1982) *Monthly Not. Roy. Astr. Soc.* **198**, 339 (Microfiche MN 198/1).

Thorne, K.S., Price, R.H., and Macdonald, D.A. (1986) editors, *Black Holes: The Membrane Paradigm* (New Haven, CT: Yale University Press).

Tolman, R.C. (1934) *Relativity, Thermodynamics, and Cosmology* (Oxford: Clarendon Press).

Tolman, R.C., Ehrenfest, P., and Podolsky, B. (1931) *Phys. Rev* **37**, 602.

Torres del Castillo, G.F. (1992) *Rev. Mex. Fís.* **38**, 484.

Trautman, A.(1956) *Bulletin de l'Académie Polonaise des Sciences, Série de sciences math., astr. et phys.* IV **665**, & 671.

Trautman, A. (1957) *Bulletin de l'Académie Polonaise des Sciences, Série de sciences math., astr. et phys.* V **721** , .

Tsoubelis, D., and Economou, A. (1988) *Gen. Relat. and Gravit.* **20**, 37.

Tsoubelis, D., Economou, A., and Stoghianidis, E. (1987) *Phys. Rev. D* **36**, 1045.

Vladimirov, Yu.S. (1982) *Reference Frames in Gravitation Theory* (Moscow: Energoizdat). In Russian.

Vladimirov, Yu.S., Mitskievich, N.V., and Horsky, J. (1987) *Space, Time, Gravitation* (Moscow: Mir). Revised English translation of a Russian edition of 1984 (Moscow: Nauka).

Wagh, S.M., and Saraykar, R.V. (1989) *Phys. Rev. D* **39**, 670.

Wallner, R.P. (1990) *Phys. Rev. D* **42**, 441.

Weinberg, S. (1996) *The Quantum Theory of Fields. Vol. I: Foundations* (Cambridge, UK: Cambridge University Press).

Wentzel, G. (1949) *Quantum Theory of Wave Fields* (New York: Interscience).

Westenholz, C. von (1986) *Differential Forms in Mathematical Physics* (Amsterdam: North-Holland).

Wheeler, J.A. (1962) *Geometrodynamics* (New York: Academic Press).

Wheeler, J.A. (1993) *Private communication*.

Yano, K. (1955) *The Theory of Lie Derivatives and Its Applications* (Amsterdam: North-Holland).

Zakharov, V.D. (1973) *Gravitational Waves in Einstein's Theory* (New York: Halsted Press).

Zel'dovich, Ya.B., and Novikov, I.D. (1971) *Theory of Gravitation and Evolution of Stars* (Moscow: Nauka). In Russian.

Zel'manov, A.L. (1956) *Doklady Akad. Nauk SSSR* **107**, 815. In Russian.

Zel'manov, A.L. (1959) *Doklady Akad. Nauk SSSR* **124**, 1030. In Russian.

Zel'manov, A.L. (1973) *Doklady Akad. Nauk SSSR* **209**, 822. In Russian.

Zel'manov, A.L. (1976) *Doklady Akad. Nauk SSSR* **227**, 78. In Russian.

Zel'manov, A.L., and Agakov, V.G. (1989) *Elements of General Theory of Relativity* (Moscow: Nauka). In Russian.

Index

acceleration of reference frame 13, 45
ADM formalism 37
Aharonov-Bohm type effect 107
analogy between gravitation and electromagnetism 58, 94, 96ff, 101
anticommutation relations 140
Ashtekar's formalism 40, 147ff

Bach brackets 4
Bach–Lanczos invariant 7
basis of a p-form 4
Bianchi identity 12
bispin 28
black-body radiation 40
Bohr–Rosenfeld 1933 paper 40

canonical
    equations
        in field theory 130
        in mechanics 124
    formalism 37, 126
    (affine) parameter 60
    variables 124ff
Cartan forms 4
Cauchy problem 51, 100
centrifugal force 39, 87
charge density
    frame-dependent 85, 89
    frame-invariant 87, 88
charges, effective (kinematic) 86, 87, 90
Christoffel symbols 9, 18
chronometric invariants 36ff, 44
classification
    of electromagnetic fields 84
    of gravitational fields 102

Codazzi equations, generalized 50, 95
coefficients $T_a|_\sigma^\tau$ 9
commutation relations 133, 138, 140
    for Fermi-Dirac fields 140
co-moving frame 62, 64
conductor (DeWitt) effect 75
connection
    coefficients 8
    forms 13
conservation laws 23, 25, 29
constraints 131, 147
coordinate transformations
    infinitesimal 10
    versus reference frame change 44
Coriolis force 39, 87
cosmological redshift 59
covariant
    derivative (,) 9
    differentiation (∇) 7
crafty identities 6, 83, 95
crossing a horizon 64
curl 48, 49
    of acceleration 85
curvature 11

de-Rhamian operator 15
DeWitt effects 75ff
differentiation operator δ 15
div curl 49
divergence 48, 85
Doppler
    effect analogue 85
    shift 62
dragging
    generalized manifestations of 76

164   Index

in the Kerr field 66
in the NUT field 69
in the pencil-of-light field 70ff
of electromagnetic fields in conductors
    and superconductors 75ff
of dragging 71
phenomenon 64
dual conjugation 5, 7, 16

Ehlers' formalism 37
Einstein's gravitational field equations 93ff
electric
    field strength 55
    type fields 84
electromagnetic field
    energy density 84
    four-potential 53
    invariants 55
    strength 54
energy (mass)
    density 111
    scalar 56
energy-momentum (stress-energy)
    pseudotensor 27, 126
    (symmetric) tensor 24
    vector 56
equation of motion, for a charged test particle 57
equivalence
    of gravity and acceleration 58
    principle (inertial and gravitational mass) 34
        relativistic revision 34
expansion scalar 46
exterior
    differentiation 12
    differential of monad 46

field strength 54
fluid 111ff
    2-form field 116
    3-form field 117
    cosmological

        constant 118
        Machian field 118, 119
        term (reinterpretation) 118
    energy-momentum tensor 23, 24
    equation of state 118, 119
    non-rotating 116
    rotating 119
    sound propagation in 121
four-current
    decomposition 85
    density 112
frequency shift 58, 64
Friedmann world 60
FW (Fermi–Walker)
    differentiation 14
        generalized 15
    curvature 14, 50

Gauss equations, generalized 50, 95
Gauss–Bonnet invariant 7
Gauss and Stokes theorems 129
general relativity, contents of 1
geodesic deviation equation 54, 93, 99
    generalized for charged masses 99
geodesic equation 14
Gödel space-time 108
    generalized 87ff
gradient of monad vector 45
gravitational instantons 148
gravitational inhomogeneity field tensors 99
gravitoelectromagnetism 75ff
gravitoelectric field 69
gravitomagnetic field 65, 69

Hamilton-Jacobi equation (generalized) 134
'heavenly spaces' 147
Hodge star 5
horizon 64
hydrodynamics 44

incoherent
    dust 112
    radiation 112

indices, collective and individual 4, 5
inertial (kinematic) forces 58, 87
infinitesimal transformations 10
invariants
    electromagnetic 55, 84
    gravitational 102
isometry 11

Kerr
    gravitational lens 67
    space-time 66
Kerr–Newman field 63
Killing
    vector field 11
    tensor 63
Killingian reference frame 64
kinematic (inertial)
    forces 58, 87
    sources 86, 87
kinemetric invariants 36
Kronecker symbol with collective indices 5

Lagrangian density
    electromagnetic 81
    gravitational 93
Lanczos identities 6
Leibniz property 8
    generalized 15
Levi-Cività axial tensor 5, 6, 42
Lie derivative 10, 11, 45, 85
    finite-difference 10
    parable of two observers 10
lightlike motion dragging 73
local stationarity 107
Lorentz force 54
    vanishing of 89
Lorenz gauge condition 82
Lovelock Lagrangians 7

Mach's principle 1
magnetic
    charges, effective 86, 91
    displacement vector 55

monopole
    dynamical 91
    kinematic 90
    type fields 84
Maxwell's equations 81, 82, 85, 86
mixed triple product 43
Møller conditions 29
momentum three-vector 56
monad 3
    basis 46
    field 41
    formalism 44

Noether
    densities 22, 28
        transformation laws 31
    relations 23
    theorem 39, 82
non-geodesic motion of a fluid 114
non-inertial frame effects 58, 87
normal congruence 45, 51
null type fields 84

Palatini approach 81, 82, 93, 148
Peierls approach to quantum commutators 134ff
pencil of light 106
perfect fluid 111ff
plane-of-orbit shift in NUT space-time 69ff
Poisson brackets 128, 131, 132, 132, 134
'polarization' 107
Poynting vector 84
pp-waves 90
precession of a gyroscope 74
pressure 112
product
    three-scalar 41
    vector 43
projector 41

quantization 37, 128, 133, 134, 148
    postulate 133
quantum physics 40
quasi-electric tensor 100ff

quasi-magnetic tensor 100ff
quasi-Maxwellian (gravitational field) equations
    new 99
    old 97

radiation-type fields 84
rate-of-strain tensor 13, 45
Raychaudhuri equation 95
redshift 58ff
reference
    body 1, 36
    frame,
        congruence of 36
        global 39, 58
        in quantum physics 40
Reissner–Nordström space-time 78
renormalization, gravitational 79
rest mass 56
Ricci
    identities 12, 13
    three-tensor 50
Riemann-Christoffel tensor 6
Rindler vacuum 40
rotation of reference frame 13, 45,
    and dragging 64

Sakurai hypothesis 120
scalar
    mass 56
    multiplication of Cartan forms 7
    product, three-dimensional (•) 41
Schrödinger-Brill formula 59
Schwarzschild space-time 60
self-dual quantities 55
shear tensor 46
signature 4
skew tensor b^a_b 42
speed of light 42
spin, generalized density of 28
spin-spin and spin-orbital interactions 74
stiff matter 112
Stress 84

stress-energy (energy-momentum) tensor 24
    electromagnetic 24, 82ff
structure
    coefficients 8
    equations
        1st 13
        2nd 13
superconductor (DeWitt) effect 75
synchronous coordinates 62

Taub-NUT field 103
τ-curvature 50
τ-differentiation ∇^τ_u 15
tetrad formalism 37
three-current density 85
three-divergence 48
three-metric tensor 42, 52
three-space 2
    curvature tensor 50
three-velocity 42
tile-roof, modification of hypersurfaces 129
transport
    Fermi-Walker 14
    parallel 14
Trautman coefficients 9
triad 37

uncertainty relation for time and energy 56

vector
    density and conservation laws 25
    product ($\times$) 43
vulnerability of the Killing vector approach 76

Weyl tensor 12, 98
    decomposition 98

Yang-Mills theory 147

Zel'manov's formalism 36
zero-point radiation 40

DATE DUE

SCI QC 173.59 .M3 R45 2005

Miδtωskevich, N. V.

Relativistic physics in arbitrary reference frames